BEE-SKEP MAKING

BEE-SKEP MAKING

HERITAGE, FOLKLORE AND HOW TO MAKE AND USE YOUR OWN SKEPS

CHRIS PARK

HERBERT PRESS
LONDON • OXFORD • NEW YORK • NEW DELHI • SYDNEY

HERBERT PRESS
Bloomsbury Publishing Plc
50 Bedford Square, London, WC1B 3DP, UK

Bloomsbury Publishing Ireland Limited,
29 Earlsfort Terrace, Dublin 2,
D02 AY28, Ireland

BLOOMSBURY, HERBERT PRESS and the Herbert Press logo are trademarks of Bloomsbury Publishing Plc

First published in Great Britain 2026

A catalogue record for this book is available from the British Library

Library of Congress Cataloguing-in-Publication data has been applied for

ISBN: HB: 978-1-78994-330-6; eBook: 978-1-78994-331-3

2 4 6 8 10 9 7 5 3 1

Typeset in Minion Pro by Jerry Goldie Graphic Design

Our printing is as environmentally friendly as possible using current technologies. We've dropped the dust jacket, abandoned unnecessary and environmentally damaging finishes, and printed the cover onto a smooth uncoated case. This book is recyclable.

Printed and bound in China by RR Donnelley
Asia Printing Solutions Ltd, Dongguan, Guangdong

MIX
Paper | Supporting responsible forestry
FSC www.fsc.org FSC® C144853

To find out more about our authors and books visit www.bloomsbury.com and sign up for our newsletters
For product safety related questions contact productsafety@bloomsbury.com

PICTURE CREDITS

Images are listed here by page number in bold, followed by the source/copyright holder:

21 Wellcome Collection/Public Domain
22 Metropolitan Museum of Art, New York/Public Domain
23 National Art Gallery of Bologna and Palazzo Pepoli Campogrande/CC BY 4.0
24 Wellcome Collection/Public Domain
25 John Paul Getty Museum/Public Domain
29 Library of Congress/Internet Archive/Public Domain
35 Falmouth Art Gallery, Cornwall, UK
36 Ceoil/CC0 1.0
38 Yale Center for British Art, Paul Mellon Collection, B1975.4.550
40 London Museum
42 Internet Archive/Souverain Grand Collège des Rites en Finlande/CC BY-NC-ND 4.0
44 WikimediaCommons/Godot 13/CC BY-SA 3.0
45 (top) Wikimedia Commons/Richard Rogerson/ CC BY-SA 2.0
45 (bottom) Wikimedia Commons/Michael Coghlan/ CCSA 2.0
46 (left and bottom) Jerry Goldie
47 (both) ©Museum of Freemasonry, London
53 Photo by Eddie Mulholland-WPA Pool/Getty Images
101 Yale Center for British Art, Paul Mellon Collection, B1977.14.11417
149 Stemonitis/CC BY-SA 2.5; **150** (top) Public Domain.

All other images are the author's own.

Page 1 image: Double hiving bee skeps (see pages 121–4).
Page 2 image: Bee skep and detail of a mead horn.
Page 3 image: Bees building wild comb out of a brood-skep.

Contents

Foreword

By Dr Nicola Bradbear, Founder of Bees for Development

One of life's great achievements and satisfactions is to make your own skep, collect a swarm of bees to make their home within it, and eventually savour some of their surplus honey, and even light a glowing candle made from their beautiful, aromatic beeswax. This marvellous book, full of treasures, can help you to achieve that wonderful aspiration.

Chris Park is one of very few people in western Europe still making beautiful, functional skeps and using them to house bees successfully in his home apiary. The instructions he now provides here are the first time that all details of skep making and skep beekeeping have been presented in this way. This must surely become a classic text: there are currently a very small number of other books about skeps – most of these are very slim volumes indeed, and for some 'pamphlet' may be a better description! It says much about the lack of written information about this wonderful craft that Chris regularly quotes from the previous outstanding skep book, written by Charles Butler in 1623, who in turn quotes from Virgil's writings on beekeeping!

Throughout his skeppist career, Chris has gleaned a unique wealth of knowledge about skeps from folklore and history, and this beautiful book begins with him sharing glimpses of the huge role that bees have played in people's lives throughout our human history: as Chris puts it, bees have been 'whispering and working away alongside us'. Here in western Europe we have probably been keeping bees in woven baskets for around 4,000 years – and it is only in the past 100 years that we have almost entirely abandoned that style of beekeeping, at the same time as we added considerable stress to honey bee populations by reducing their habitat and poisoning them with agrochemicals.

Because honey has long been the major source of sweet food and medicine for humans – a wonderfully useful, concentrated substance that would never go mouldy, or could be transformed into alcohol – this means that our culture, folklore, history and languages are rich with honey-full references.

For the beginner skep maker, two aspects can be tricky: the first is, if you are planning a straw skep, to locate a supply of suitably long straw – modern wheat varieties are too short. Secondly there is getting going to create the skep from lipwork. Skep making begins at the crown, with a tight spiral of straw, and getting this right is one of the great secrets to a very good, strong and long-lasting skep

– Chris has now captured in pictures, and describes in words, every step of this process – these are given in Chapter 4, along with detailed pictures and instructions of each stage of skep making.

A skep alone does not constitute a complete home for bees: like us, bees' housing requirements include an entrance that will ensure security from unwanted visitors, dressed walls (i.e. cloomed like a cob cottage), a steady and dry floor, and shelter from the elements – heat, cold and rain. Chris shows exactly how to make all of these and the various options available to you – chapter 5 is devoted to making hackles – conical hats made of straw, reed, or willow, and the other ways to provide suitable shelters for skeps.

Chapter 6 is an exceptionally useful and detailed guide to skep beekeeping. Like skep making, beekeeping in skeps is nowadays little known and practised in western Europe, whereas at the start of the twentieth century these skills were commonplace. In industrialised nations like the UK, frame hives were introduced by the government as a way to standardise beekeeping: by using modular hives, colonies of bees become transportable, saleable and accessible for people to manipulate as they desire. When new technology is introduced, it is our human behaviour to rubbish what had gone before – and this is what happened to skep beekeeping. Even now, novice beekeepers are still told the old chestnut that in those far-off and backward days, 'skep beekeepers killed the bees to get the honey'. This would be akin to killing the cow to get the milk! Of course, that was not the common practise – and in Chapter 7 you will find details of how harvesting honey from skeps was (and still is) done – one method involves the temporary driving of bees from their home skep into an empty one, so that the beekeeper can access and harvest some honey combs. Indeed as recently as the 1980s there was a competitive class for 'driving bees' held within the Mid Somerset agricultural show held in Shepton Mallet every third Sunday of August. With the help of this fabulous book, maybe we can hope that this class may one day be restored!

Chapters 8 and 9 are devoted to showing how to process honey and beeswax from skeps, with Chris sharing his important secrets of mead and metheglin making – I can testify that he is certainly an expert in this!

I very much appreciate all the skeppish knowledge and know-how that I have gained from Chris over the years – and am delighted that through this book he now enables many more people, and indeed future generations, to discover the wonderful art and craft of skep beekeeping.

Nicola Bradbear MBE
Founder of Bees for Development
Monmouth, October 2025

Introduction

Now what delight can greater be
Than secrets for to knowe
Of sacred Bees, the Muses' Birds,
All which this booke doth show

CHARLES BUTLER,
THE FEMININE MONARCHIE, 1609

This book is about bee skeps, but is also about the virtues of honey bees, exploring some of the ancient lore of the Isles of Britain, Ireland and beyond. At times it traverses comparative mythologies and attempts to reveal some of the science within the sorcery of it all. Along the way one will find reminders and remnants of traditions through which honey and other bee produce have been ever-present substances; where the heritage of honey has sweetly and subtly fortified our rites of passage, our language, and the most meaningful moments of life, from cradle to grave.

Similarly, this book muses upon the milestones of human history, where bees, their medicines, and symbols, have been whispering and working away alongside us, as if driving and guiding our evolution. Honey bees and bee products have traditionally been woven into birth rites and first rites, betrothal and marriage, sex and sexuality, family planning, fertility and contraception, health and happiness, art and music, clothing and community, mysteries and cosmologies, religion and spirituality, chastity and piety, kink and pleasure, medicine and warfare, palliative care, and last rites.

Honey bees are thought to have descended from giant carnivorous wasps who became vegan, grew their hair long, and began to live in full-time communes. This happened long before the 1960s of course, in a gradual fashion, according to the natural pace and the subtle patterns of evolution, but those early wasps had garnered a taste for tree pollen, which was predominantly wind-blown way

Sunlight shining through honey bees swarming on a tree trunk.

back then, long before the sensational aeons of more mellifluous flora. Trees and plants would release billions of fertile grains to float and fly as microscopic dust, which many species still do to this day from cones and catkins. Pollen is extremely nutritious and contains more protein per gram than beef. These ancient wasps, and some of their coleopteran cousins, were on to something.

Over millions of years, plants and trees slowly responded to these beneficial pollination services and began to evolve accordingly, with such magnanimous magnificence that the world hasn't been the same since. Flower power ensued in full force as planet earth bloomed into a sweeter, more colourful and aromatic place. The plant world developed rich nectar as a nutritional carbohydrate reward for their pollinating friends, and fragrant petals as a beacon and lure.

Bee biology

To start with, I will offer some basic biology and science here and will endeavour to give you more throughout the book.

A healthy queen bee has two internal bags of eggs, her ovaries, and will store sperm collected from the drones she has mated with close by in another special container in her abdomen called a spermatheca. When laying an egg, if she wishes it to develop into a female worker bee, she will pump some sperm from her store to meet it. This will become a daughter who will have 32 chromosomes.

If she is laying into a larger drone cell, she will deposit an unfertilised egg. This will develop into a male bee. He will have a grandfather, no father, and only 16 chromosomes. The sole function of the drone bees we are told is to pursue queens and mate with them. They also sing loudly, which may be good for the morale of the colony in summer months, but not in early autumn when resources are precious and they are evicted by the worker bees. A drone may have other ways to serve the colony of which we are not yet aware. If he is successful in catching a queen to mate with, he will die in an orgasmic ecstasis as his endophallus explodes on ejaculation, with an audible 'pop!' The mighty widow-queen flies on, sometimes with the drone's reproductive organs still trailing behind her.

Most of a bee colony consists of the female worker bees, who do not have a fully developed reproductive system, and do not have sex, but do have eggs. They can lay them too, in an emergency, if the colony becomes queenless. These eggs with the single set of chromosomes will develop into haploid drones as above. Hundreds may emerge from a queenless colony after being diligently nursed and fed. This is a last-resort effort by the workers to make the

A call to arms

Bizarrely, bananas have been found to rile bees. They contain isoamyl acetate, one of the chemicals in their alarm pheromone. When worker bees feel threatened, they will release this pheromone from their Koschevnikov gland, near the sting shaft. It is also released when they sting another animal or when they stick out their stinger. The scent attracts other bees to the location and causes them to behave defensively.

most of their resources and send their genetic heritage and wisdom out into the bee world.

It can be appealing to our human nature to view a colony of bees as a complete organism: each bee being a cell in this superorganism, the queens and drones the reproductive system, the workers all the other metabolic functions, and a skep hive the organism's clothing, or skin that it sheds from time to time. The one large bee in the middle of the colony has been used by religions, states, and authorities to condone autocracy as model society – the 'one king', 'one pope', 'one God', periods in history. We now know, however, that the large bee is not only a queen (not a king) but neither is she a manager nor an autocrat of any kind, living in service to the whole and having less freedom than the workers and drones. A colony of bees is an autonomous collective of individuals able to make their own decisions according to all the factors they are aware of, communicating continually when together, and when alone and on the wing, continuously devoted to the whole.

Bees and humans

Humans swaggered onto the scene long after bees were an established species. These primal honey hunters developed their ancestral skills and passed them down through the many generations of bee botherers until our recent relatives were sharpening honey-poking sticks, making ropes and ladders, and curing hides for honey bags, all of which helped secure some success in the stealing of sweetness and medicines from the bees. A bond developed between the eusocial insects and the prosocial humans.

Wherever one finds honey bees, we humans have feasted upon their sweetness and made offerings of their bounty as if it were our own. Throughout the ages, whilst supping upon the wholesome honeyed brews that have provoked great insight, prophecy, and progress, we have revered honey bees as among the most high and sacred of beings. They have been upheld as creatures of the gods, who bestow great blessings upon the world, inspire cultures, civilisations, and customs alike. Honey bees appear to have a mellifluous and mysterious history woven through our own. From the dawn of time, they have been a source of sustenance, health, and fertility for many species, and within human history a foundation of rich symbolism and divine inspiration.

> *Ubi apis ibi salus.*
> Where there are bees, there is health.
>
> Attrib. Pliny the Elder (23/24–79 CE)

This interspecies connection progressed on a more practical level to become what we are familiar with today as beekeeping, and it has diversified greatly. Over thousands of years many cultures developed apicultural practices to give honey bees a home, using handmade vessels crafted from whatever was to hand. People fashioned hives from grasses and hollow logs, tree bark, earth, and ceramics. Whether made from plastered wickerwork or plaited leaves, all were tailored to the needs of the honey-bee colony and the wants of the beekeeper. It may surprise you to know that most of the honey bees that are 'kept' in the world today are still kept in these indigenous ways. It surprised me when I learnt that, because the conventional wooden-box beehives with moveable frames are so prevalent where I am, in the UK.

Beekeeping history

Where and how did the practice begin within the British Isles? Archaeologists are undecided on a date, and even the era, in which we stopped 'honey hunting' and started 'keeping' bees in this neck of the woods. There is no hard evidence for beekeeping here before the Romano-British period because baskets and logs rarely turn up in the archaeological record. Julius Caesar noted that British natives were most proficient in all types of basket work. Before Roman settlements, folks may have been managing wild colonies in trees, perhaps also log-hives, perhaps wicker-hives. It is pertinent to note that there was nothing written or recorded to say that the ancient Britons and Irish weren't keeping bees, which would have been noteworthy to those writing about the indigenous cultures here.

I like to think that we were beginning to keep bees in the UK and Ireland in woven baskets and log-hives as far back as the late Neolithic Age, around 4,000 years ago. Colonies would have originated from swarms attracted to some kind of vessel that may have been used to contain wax, honey or other bee produce, perhaps as a received praxis from further afield. It is very tempting and too easy perhaps to imagine a jump from honey hunting to skep beekeeping happening through a colony of bees swarming into a basket or vessel, when honey hunting and tree beekeeping were gradually replaced by log-hive and skep-hive beekeeping, and the hive enclosures and bee gardens were born. A new relationship with bees began as we started better to understand their language and behaviour to ensure a harvest.

Worldwide, these honey thefts became more cunning and well-crafted as refined techniques and taboos developed into social codes and sacred rites, trade secrets, and cultural laws. The Bronze Age tin-trading link forged with the Phoenicians, who were accomplished beekeepers, inspired themselves by the Minoan culture, may have influenced apicultural advances here. The tangible, malleable, and pertinent link between the developing metalwork technologies of the Bronze Age and honey bees is beeswax. The process of 'lost wax' was used in casting objects from gold, silver, and bronze. As an old saying teaches us: 'If you want to make something in metal, you must first make it in wax'.

Thousands of summers later, European forest beekeepers were managing wild colonies high up inside hollow trees, cutting out rear doors to enable access to the comb and leaving their personal mark upon the trunk. As the ages progressed, this evolved into collating the colonies within special forest and heathland clearings, fenced or walled off from bears, deer, and pigs, containing either hollow log-hives or bark-hives stood on end. The hives were sometimes roofed with thatch, wood, or stone/slate. Fine examples are still in existence and use, most notably within the Lithuanian Museum of Ancient Beekeeping.

A wooded clearing or a heathland enclosure for bees, often found on south-facing slopes, was originally called a *garth* or a *garn* in parts of Britain. This is thought to be the root of the word 'garden', being *gardd* in Welsh. Before the more scientific term apiary became popular for the place where one kept honey-bee colonies, it was simply called a 'bee garden'. In northern Spain, one finds the stone-walled *albariza* enclosures (seemingly relating to Virgil's *Alvearium*); on the heathlands of Germany, the *bienenzaun*. In northern India, whole royal forests were managed solely for honey production, known as *madhuban* or *madhuvan* – literally 'honey-forests'.

What is a skep?

The most well-known indigenous hive within the Isles of Britain and Ireland, and throughout much of Europe, is a kind of basket most often called a 'skep' today in the English language. This word is rooted in the Anglo-Saxon word *skeppa*, which simply means 'basket'. There you still have it, implied in the second syllable of basket. The name skep is cognate with many others that denote vessels and containers: cup, cap, cape, scoop, ship, skip, keg, kettle, coop, keep, and keeper. I enjoy this etymological thread linking bee skeps and beekeepers. In Scotland, however, a skep is often called a *ruskie*. Again, this is a name for a basket and may well have the same origin as the French word for the beehive, *la ruche*. Some scholars believe this has its root somewhere in the use of rushes, and others suggest it is in the use of bark. Both are commonly utilised in basket and beehive making across the world.

Honey etymology

'Mmmm', we sometimes exclaim when eating something tasty like honey; and sounds preceded language. Most of the Indo-European derived words for honey are 'm' words, their etymological root being the Proto-Indo-European *méd*^{h}*u*, and the Sanskrit word *madhu* – still present in Indian culture today and meaning, of course, honey. The Proto-Celtic names became the Gaulish *medu*, Welsh *medd* and *mêl*, Cornish *medh*, Breton *mez*, Old Irish *mid* and Gaelic *mil*. Consider the Slavic *med*, the Greek *meli*, the ancient Hittie *melit* and the modern French *miel*, all being akin to the Chinese words *mih* and *mat*.

A display at a National Honey Show (UK) bee-skep-making workshop.

The straw bee-skep-making technique is known as 'lipwork' or 'lippwork', whereby straw ropes (of wheat, barley, or other grasses) are bound into shape using thin laps of bark (willow, holly, bramble, etc.) or string in the production of chairs, beehives, and other baskets. I'll wager it's so-called because of the use of lapping in the process to bind the straw. I wouldn't be surprised if it has been known as 'lap-work' at some point in history, and may relate to the word 'lath', a long thin slat of wood. Another theory is that the name derives from a 'seed-lipp', a kind of basket carried by farmers carrying out the broadcast method of sowing seed, but it could well be the other way round: that the seed-lipp is so named because it is made of lipwork.

Skep making today

Skep making is currently an endangered craft. Several years ago, I received an email from the then Heritage Crafts Association (HCA, now just 'Heritage Crafts'), asking me a thorough list of questions to help them assess where skep making sits on their list of waning crafts. I diligently filled in the form, naming all the makers and teachers I was aware of, estimating the numbers of annual courses and classes in the country, publications, and more, thinking that it wouldn't be classed as endangered because many beekeepers attend many courses throughout the year. It is indeed a popular activity amongst beekeepers. However, no one is a full-time bee-skep maker within the UK, nor far beyond. They take a long time to make, and it just doesn't pay. A couple of the makers and experts I had originally listed for the HCA have now sadly passed away, the body of practised teachers is diminishing, and you never know when your number's up. It is important to say that the levels of expertise, skill, and competence of skep-making teachers can vary greatly, but all pass on great enthusiasm, and we improve with each coil of the straws.

Those of us that are familiar with bee skeps are so because we are beekeepers, know a beekeeper, or are involved with country crafts somehow. They have mostly survived into contemporary beekeeping practice in the UK and Ireland as swarm-collection vessels, rather than permanent hives. When a swarm of bees has settled onto a tree, in merry weather, full of purpose, there is nothing finer than a well-wrought skep to ease underneath and shake them into. Most of the students that I teach to make skeps have this very moment clearly envisioned in the back of their minds whilst diligently working away. Times are a-changin' though, and there is a gentle resurgence of the craft of skep making for the purposes of skep beekeeping.

I'm often pleasantly surprised at how popular the skep-making workshops I run are, as you cannot easily make one in a day. In fact, a student on a one-day course will make approximately one third of a skep if they do well. They then take away the materials, tools, and the knowledge to finish it off at home. Those that make more than a third have usually rushed it and could have made it a bit stronger, and I didn't manage to slow them down enough. These things take time, and the more you put in, the more you get out. This book I hope will be of significant help to any budding student, and of interest to seasoned beekeepers and those fascinated with the bee world.

I am incredibly pleased that the craft and the practice of skep beekeeping are having a renaissance, along with a general interest in bees, their medicines, and the craft of mead making. It can be a salubrious and sensual art, ensuring healthier and happier colonies of honey bees, and in turn, a healthier environment, and more contented people.

Skep beekeeping – learning by experience

From looking at old bee books and manuscripts, such as the writings of the Revd Charles Butler (see Chapter 3), it is clear to see that every bee-master had different preferences and practices, varying their styles of management according to their situation, wants, and wiles. As a researcher and reviver of skep beekeeping and a teacher of skep making, I have not yet standardised my own practice. My way of learning is to be continually experimenting with a range of styles and shapes, seeking to experience as much as possible. For your benefit I'll try to be more focused here, concentrating on a few simple steps and methods. I do intend to impart some expert and comprehensive instruction on creating a strong skep that will have longevity, and then to suggest how your craft might vary. You can expect some guidance on a couple of ways to keep bees in skeps. I'll point you towards the older sources if you wish to become as niche and nerdy as one can.

Looking to the future

When someone expresses a concern to me about the craft of beekeeping or how to look after bees, I'll habitually answer them with something encouraging like, 'Try not to worry too much, they've been looking after themselves for well over 40 million years, as we know them'. I may reassuringly add, 'They simply need a good home, with protection from the elements, pests, and predators'. If someone enquires more, I might start talking about the bees' ancient wasp ancestors mentioned above, who were hunting and stinging their prey to feed themselves and their young over 120 million years ago. I'll revel again (there is a lot to revel at surrounding bees), at that mind-boggling amount of time, throughout which all bees and many other insects have been evolving strategies to combat and cope with viruses, predators, changing environments, mass-extinction events, and whatever problems needed solving along the way. Compared to this time-tested species and their wealth of ancestral experience, genetic heritage, and success, and what many cultures have considered to be divine wisdom, we are a toddler of a species, still learning to walk, crashing into things as we go. One might say that bees have learnt to eight-time waltz around the world's dancefloor whilst we are playing with toys and running anthropocentrically about causing chaos, only just realising the consequences of our actions. Bees are experts at adaptation and cunning problem solvers. A pertinent question might be, will they adapt to our world-changing manipulations and miticides that we unswervingly meter out regardless of any wise-ape ways and science, portents, and sense?

In 1951 Karl von Frisch wrote that bees' lives are like magic wells; that the more you take from them, the more they fill with water. I'm in two minds, and hearts, about this. A part of me nods and smiles and knows that once a person begins to be inspired by bees and 'gets the bug', a lifetime of unfathomable and salubrious inspiration ensues, including the inevitable stings. Bees offer so much to so many – artists, scientists and naturalists alike. Von Frisch was a brilliant scientist who discovered and affirmed much of what we know today about honey-bee

Various styles of skep on show at the Bristol University Bees & Pollinators Festival.

communication, perception, and behaviour. His excitement about and veneration of the honey bee is contagious.

Yet, there is another part of me that reacts in a different way, concerning natural resources. I know this isn't what von Frisch was implying, and that he's using a fairytale allegory, and not talking about resources. Nevertheless, I am triggered by the notion of an 'unending supply' that can promote a kind of greed; and we might be in danger of taking these supplies for granted. As a species, we tend to feast on resources until they are gone. We have drawn and drawn, and over-abstracted, and abused the generosity of many wells for our own ends, trying to market and patent, genetically modify and manipulate the magic. How might we progress with less anthropocentric appetite and more respect? What gratitude might be offered to this miraculous source, and what might aid a more careful relationship, so that the spell is not broken?

Chapter 1

Myth, Folklore and Superstition

In the rituals of the worship of the goddess:
Let the offering of honeycomb be washed
With milk and with soft wine

Virgil (70–19 BCE), Georgics (Book I)

This chapter is a saunter through apian folklore and mythology to extoll their virtues and venoms, exploring the heritage of honey.

African traditions

Some of our most ancient human cultures revere the selflessness of bees in their creation myths. The skilful San people, formerly known as Kalahari 'Bushmen', are part of a 100,000 year-old culture that survives to this day. Their elders and cultural teachers speak of two beings, Bee and Mantis, who existed in an aeon of great deluge when the world was covered with water. In this beginning, all of the deities lived in the Sky. Flame created the earth, and Rain was his mate, who carried the rainbow as a girdle around her waist. Mantis, a kind of demigod figure, was sent to find the purpose of all life and to ask Bee to guide him. Bee offered Mantis a ride across the primordial waters, being a more accomplished flier and a born navigator. After many days of exploration Bee became weary and cold, searching for solid ground. Mantis felt heavier and heavier as Bee struggled on, flying more slowly and sinking towards the water. Then, something came into view, growing upon the surface of the flood: a white flower. Bee pushed on with a last effort.

A mystical assortment of cloomed skeps and hackles, with a hen house behind, made for use in the film *Hamnet*.

Before dying of exhaustion, and carefully laying its passenger into the heart of the flower, Bee put the seed of the first human being inside Mantis. The sun warmed, encouraged, and empowered the flower. Then Mantis awoke. From within this primordial being, from the seed left by Bee, the first of the San people was born. In other tales, Mantis, also known as Ikággen, was the husband of Ihúnntu!att!atteȩn, who was called the 'Mother of Bees'.

In contemporary West Africa, the Bambara people are around five million strong. Although converted to Islam, they still adhere to some of their ancestral traditions, affording themselves a customary cup of mead, which they say bestows three things upon humankind: health, knowledge, and the truth. They also declare the truth is like a honeycomb, having neither a wrong side nor a right side, and is the sweetest thing in the world.

In Morocco, bees and honey are said to contain a potent amount of *baraka*, meaning they are full of 'divine blessings' and the 'beneficent force' that flows through all physical and spiritual spheres. They also say that if someone dreams of a bee entering a person's mouth and flying out again, they will become a good singer.

Hindu traditions

When all other deities had failed to defeat the power-seeking titan Aruna because he was invincible to man, woman, two-legged or four-legged creatures, the supreme 'mother of the universe', Adi Shakti or Mahadevi, issued forth in this time of great need as Bhramari, the goddess of black bees, wasps and hornets, whose six-legged nature saved the day.

Several Hindu deities are represented by bees, enclosed or accompanied by them, and are ritually offered honey. A single bee atop a triangle signifies Madheri, the suave and less destructive form of Lord Shiva. The three gods Vishnu, Krishna, and Indra are collectively known as the Madhava, the 'nectar-born' gods. Lord Vishnu is sometimes depicted as a blue bee on a lotus flower, where the blue represents the sky, the world of the gods. The lotus is a symbol of feminine beauty, prosperity, and fertility (I sense a comparative mythology emerging here with the San creation myth mentioned above). The Hindu deity of protection and compassionate love, Lord Krishna, being the eighth avatar of Vishnu, is often depicted with a blue bee on his forehead and is often portrayed with blue skin.

In Hindu mythology, the deity Surya, a solar god, has two horse-headed twin sons known as the Aśvins. Every morning as lords of light they ride over the horizon in their chariots laden with mead, with their whips bestowing honeyed blessings and strength. A line from the Sanskrit hymns of the Rig Veda runs 'O Aśvins, lords of Brightness, anoint me with honey of the bee, that I may speak forceful speech among men'.

A bee presents honeycomb to the Olympian gods. One story tells of Zeus granting bees a sting in order to protect their honey, but with the downside of dying having used it. Etching by Francis Barlow (1622–1704), 1668.

The all-pervasive Kama, or Kamadeva, is the Hindu God of erotic love and desire. His bow is made of a sugar cane, and his bowstring is a chain of bees. He carries honey-soaked weapons to unite the hearts of couples and shoots the five floral arrows of desire, each with a different effect: desire, attachment, passion, infatuation, and joy. Rati, the Hindu goddess of carnal pleasure and desire, is Kama's counterpart and is often seen firing the unavoidable arrow-flowers from the same bee-strung sugarcane bow.

Greek myths

The Greek god of love, Eros, dips his passion-inducing arrows into either honey or venom before shooting them, prescribing for his victims the medicine he thinks they need for that particular love-struck episode. Known as Cupid in Roman mythology, he is often seen as a moody meddler in human relationships, whose anointed darts are said to produce 'love at first sight' or the 'madness of the gods'. Bees are sacred to Eros's mother, the goddess Aphrodite (Venus), and there are several artworks that depict Aphrodite/Venus and Eros/Cupid with bees and honeycomb. When he was stung by a bee while stealing a honeycomb, his mother reminded him that he inflicted far greater wounds himself.

Melliseus (Meli-Zeus) is an ancient Greek 'honey-man' or 'bee-god' and father of two nymphs who nursed the infant Zeus on Crete. He was the leader of the dancing Corybantes warriors who guarded the baby god as he fed on milk and honey.

Within many myths, honey and honey products are used in the placating and calming of powerful forces. This is exemplified in Greek mythology by the lulling to sleep of Cerberus, or Kerberos, a guardian of the gates to the underworld. This mighty being was put to sleep with honey-cakes, or barley-cakes soaked in mead, so that a hero or heroine could gain entrance to the realms of the dead. This in turn influenced some funerary customs of ancient Greece in which a honey-cake was placed with the deceased to aid their descent. Often this was accompanied by 'Charon's obol', a coin for the ferryman, placed within or upon one's mouth.

Bugonia: honey from ox carcasses

Venus with Cupid the Honey Thief, **copy after Cranach the Elder, *c.* 1580–1620.**

From death, comes new life: being to non-being, and to being again. The myth of the Greek god Aristaeus and his bees is rooted in an ancient Egyptian tale. He was raised and tutored by many gods, including the mysterious Thriae (who nourished him on nectar and ambrosia). He becomes a master and patron of many rural and rustic crafts, including beekeeping, and brewing, but then loses his bees as a punishment, and is assisted by his mother Cyrene, who instructs him in a ritual. He duly makes a bovine sacrifice and inside the carcass of the bull, a swarm of bees appears, and he once again becomes a beekeeper.

In one Egyptian version of this 'ox-borne bees' practice, the bull is buried with its two horns protruding above the ground. They are sawn off and the swarm issues from their two hollow roots. This is a wider motif in mythology known as bugonia. Of course, we are all thinking that a carcass of such an animal will produce a swarm of flies, not bees. Yet the symbolism here is of sweetness emerging from strength, and the teaching may be allegorical, esoteric, celestial and nautical. The same motif appears in the story of samson and the lion, 'from strength comes sweetness'.

Roman myths

Many ancient Roman myths were perpetuated by 'mystery schools' to engrain knowledge and wisdom into initiates, including much star-lore. Navigation and seasonal reckoning by the stars were paramount. The bugonia motif indicates to us the constellation of Taurus, which contains the clustered asterism the Pleiades (or Seven Sisters) within its shoulder – stars that disappear into the west in late spring and are 'reborn' in the east in Autumn. Virgil (70–19 BCE) writes of the widespread belief that the Seven Sisters returning to the night sky tell us it is time to harvest honey.

Engraving depicting ox-borne bees, or bugonia, by Coriolanus Giovanni Battista (1590–1649).

At dissolution! There is no death: but they live and they fly,
To the ranks of star land and enter the high reared hall
of the sky!

Virgil (70–19 BCE), '*Epic of the Bees*' Georgics (Book IV)

Some spiritual schools talk of the moon as a hive, and the stars as honey bees. In antiquity, it was thought that chaste bees collected their young from flowers, and honey came from heaven. Therefore, the characteristics of harvested honey were often put down to the movement and influence of the stars. Pliny the Elder, in his Natural History, Book XI, chapter XII, affectionately calls honey the 'saliva that comes from the stars' being 'engendered from the air, mostly at the rising of the constellations, and more especially when Sirius is shining'.

There was a belief that good and bad honey were produced when certain signs were in the ascendant, the best being gathered when Sirius ruled the skies in conjunction with the rising of Venus, Jupiter, or Mercury. This honey was deemed a powerful medicine for many ailments with a reputation for reviving the dead.

Mellon or Mellonia (whose name derives from the Latin 'mel' for honey) is a Roman Goddess who takes care of bees and protects the sweetness of their honey, and was of great importance to Roman beekeepers.

The Roman bull/ego-slaying god Mithras also deserves a mention. Of Indo-Persian origin, his cult in and around the ancient Roman world involved solar and bee symbolism and the ritual use of honey. In the ancient mystery school, a candidate progressed through seven initiatory planetary degrees. The name of the second degree, Nymphus, is symbolic of a bee-chrysalis, and under the auspices of Venus, the veiled apprentice undergoes a period of celibacy and incubation. Whilst working through the latter degrees to attain the maturity of an adult bee, the candidates are washed with honey, at one point having their tongue anointed with it to cleanse their speech of all guilt.

In this engraving of a Roman harvest celebration by B. Picart (*c.* 1733), one can witness the importance of skep beekeeping within the scene, and figures beneath the tree on the left can be seen using the practice of 'tanging' (see page 101).

British myths and superstitions

The heritage of honey and honey drinks within northern European shores is substantial. There is evidence on a Neolithic pot-shard deposited around 4,000 years ago of such drinks being enjoyed in Scotland. Pictish brews of fermented honey and heather were renowned for being intoxicating, healing, and calming; they were coveted before agriculture raised the bar and the barley, and other grains that were added to create honey beers.

In one of the Celtic 'Otherworlds', a stream of mead forever flows: 'Rivers pour forth a stream of honey, In the land of Manannan son of Ler'. Manannan is the Irish god of the seas, and the crests of the waves are his horses. The Isle of Man is named after him. Upon the sacred Irish hill of Tara, where the high kings would gather, there once stood the 'house of mead-circling', a large banqueting hall.

Honey drinks and meads were potently omnipotent in the British Isles. Mead is the drink of the gods, the beverage of kings, the dainty drink of the gentry and the elixir of immortality. It is abundantly celebrated throughout many of the traditional stories, songs, and poems of Britain. Some say that Ceridwen, the Brythonic Goddess and mother of poetry, inspiration and wisdom, was fermenting honey and grain in her magic cauldron, heralding developments in the craft of brewing. If one looks back far enough, the names for mead and honey are the same, *Med.*

One Welsh Triad preserved by Geoffrey of Monmouth in 1136 tells us that this verdant land of Britain, when first ever settled, was known as Y Vel Ynys – 'The Honey Isle'. Among the many advantages he describes are the 'flowers of many colours that do proffer their honey unto the bees that flit ever busily about them'. If the name 'The Honey Isle' is interpreted literally, one can imagine a land mass blessed with an abundance of honey bees and their sweet produce. Apiculturally, one might assume it was home to tribes that held honey, honey hunting and/or beekeeping at the core of their identities.

Britain was also referred to as Ynys Fel Veli, 'The Honey Isle of Beli' – a Celtic mythological founder-king/god – in a medieval Welsh poem said to have been composed by the prolific bard Taliesin for the inauguration of King Arthur.

Another triad lists 'the three plagues of Britain' during King Beli Mawr's time, one of which was a horrific, debilitating shout heard over every hearth in the land upon magical May-eve. The sound came from shape-shifting creatures that would meet at the very centre of Britain at this time to wrestle. They would become dragons, one red and one white, emit the terrible screech during battle and finally descend to the earth as small pigs. Harmony and health were restored to the land when Beli Mawr's son and heir King Lludd, under the guidance of his wise brother Llefelys, calculated the centre point of Britain as Oxford, dug a pit and filled it with mead, which the creatures drank and fell asleep. They were then wrapped in a brocaded silk sheet, secured into a stone casket and interred beneath a strong mountain.

A depiction of bees gathering nectar and returning to a skep hive, a detail from the 'Northumberland Bestiary', *c.*1250–60.

Beekeeping superstitions

There are many superstitions in the craft of beekeeping, and I like to adhere to some of them; you know, just in case the mead sours or something…

To name a few, it is considered bad luck to buy bees, except with the purest gold. It is also unlucky to sell them, but some say they may be bartered for with goods of equal value, like a few bushels of corn or a large pig.

> *If you wish your bees to thrive,*
> *Gold must be paid for every hive;*
> *For when they're bought with other money,*
> *There will neither be swarm nor honey*
>
> TRADITIONAL SUSSEX SAYING

Once you are looking after them, it is unlucky to count your hives. If you do, they might perish or become diseased. Some beekeepers used to arrange their hives in a way that made it difficult to count them. I must confess to having colonies well spread out of sight through a young woodland, in many different styles of hive. Today, standardisation can be helpful for beekeepers who like to use a particular dimension of frame or hive. With skeps and fixed-comb beekeeping one doesn't have to measure too precisely. In the warmer months, some colonies will be about to swarm, and some will be newly settled swarms. The number of colonies rises in summer and can fall in autumn due to queens not mating properly, and other factors. People often ask me how many colonies I have and without fail I answer,

It is considered good luck for a bee to land on you.

with a twinkle in my eye, and sometimes tapping the end of my nose with my index finger, 'it's unlucky to count… unless you're an accountant'.

It is generally considered good luck if a bee or bees arrive at one's home. If one lands on your hand this is thought to bring especial luck, and if it lands upon your head a degree of success is to be expected. I have mentioned already how the magic of bees is associated with eloquence, and how, in many of these stories, they appear as an omen at the birth of poets and orators. There is a belief that a queen bee will settle upon the lips of a favoured person.

The luck of swarms, however, swings both ways. If they alight and cluster onto living wood, it is a very good omen. But if they do so on dead wood, it is said to predict a death.

There is a belief that humble beekeepers and their bees have a spiritual connection, as if their souls have become woven together somehow. There is a tradition of telling the bees news, as if they were family members, or 'winged messengers to the gods'. They must be kept informed of one's sorrows and joys, or they will wane away within a year and a day. Any news must be relayed to them gently and politely.

A lesser-known belief is that bees are offended by quarrelling and will never prosper if looked after by a tempestuous family. In fact, they are thought to dislike all forms of bad manners, impoliteness, and uncleanliness. Some say they appreciate chastity before a visit and are also thought to be offended by strong body scents, sweat, menstruation, spicy or malodorous breath, and other such emanations of the hairy mammal. Bee suits and protective clothing are customarily pale-coloured or white, which is believed to be calming, whilst stronger colours are believed to trigger their atavistic 'bear!' response.

Bees will be unusually aggressive when a rainstorm is brewing, or thunder and lightning are looming. The weather-lore expert Richard Inwards wrote in 1893, 'When bees to distance wing their flight, days are warm, and skies are bright. / But when their flight ends near their home, stormy weather is sure to come'.

There is a superstition, perhaps allegorical, that if a person removes stinging nettles from around a beehive, then they will duly receive the sting of the bee. Perhaps reminding us that acting out of fear of something can often bring on the very thing we are fearing.

Bees have had a continuous presence in witchcraft, sorcery, and high and low magic alike. There are stories of cunning folk and turnskins shapeshifting into bees or having bee helpers, with similar 'evidence' from Scottish witch trials. Even though these 'confessions' were the desperate result of abuse and torture, a pre-existent folk belief is there.

Bees in the human life cycle

Bees and their magic have been seamlessly intertwined with every rite of passage and milestone of human existence.

Marriage rites

Marriage customs worldwide are steeped in bee lore, honey and mead. In some leafy parts not so long ago, young newlyweds and guests would sing a blessing song at weddings: 'diligent is the life on a farm, like the life of the bee, and marriage is sweet as honey'. In Scotland, there was even a tradition of inviting bees to a wedding, and across many cultures telling them all the news of course:

> Marriage, birth or buryin',
> News across the seas,
> All you're sad or merry in,
> you must tell the Bees.

Rudyard Kipling, 'The Bee-Boy's Song',
Puck of Pook's Hill, 1906

Beehives in Brittany were decorated with scarlet cloth to mark a nuptial occasion, whilst in other parts of France a white ribbon was used. In Persia, couples would feed each other honey before exchanging vows and the first kiss. In ancient Egypt, some grooms would promise to provide their wives annually with 32 pounds of honey. Within Hindu culture, a groom may sometimes announce to his bride something like 'This is honey, the speech of my tongue is honey, the honey of the bee is dwelling in my mouth and in my teeth dwells peace'. In many parts of Europe, a hive full of bees was a thoughtful wedding present. Most beekeepers today will tell you that the term 'Honeymoon' is derived from an ancient custom of gifting newlyweds a month's worth of mead or honey, to aid fertility and conception.

Aphrodisiac and fertility

Before every beginning, every creation, or alchemical rebirth, there is often a bit of love-making. The realms of mythology, story, idiom, folklore, and metaphor tell us much about what it is to be human. If a San person asks, 'have you eaten honey today?', they are actually asking about your sex life. Honey is indeed a famed aphrodisiac, and has been found to boost sex drive, increase testosterone and oestrogen levels and swell sperm counts. It can be effective against many maladies of the loins, from fungal infections to weak erections. The Ancient Greeks mixed it with pepper to make a potent elixir that was used as the preparatory, spicy Viagra of its time. Arab cultures added onion juice to similar stimulating effect.

Richard Dagley, 'Taking caudle', 1821.

Wherever one looks, honey, honey bees, and their prosperous potency have been associated with family planning – notably with aiding fertility and conception, but on the odd occasion with contraception too; it all depends on where you put it… There is a gynaecological papyrus from ancient Egypt that describes the effective use of honey mixed with powdered crocodile dung as a kind of contraceptive paste. Being spermicidal and mixed with the dung, it creates an antimicrobial barrier preventing spermatozoa from swimming along fallopian tubes.

In Asia, eggs mixed with honey are commonly used for a variety of sexual health issues affecting men and women. This 'golden elixir' is included in massage oils for treating erectile dysfunction and employed during the postpartum and post-menopausal periods to improve health and enhance vitality. A long golden thread of powerful 'apitherapy' (honey-bee-medicine) stretches back through history.

First rites

After the marriage, mating, and ever-changing consummation customs of one's culture, if it is the will of the betrothed and the gods, gestation and birth may be the next rites of passage. In Europe, honey has been an ever-present part of a traditional midwife's medicine cabinet. There is a lesser-known, hot and spicy honey beverage named 'caudle' that used to be given to mothers in their bed-rest after labour, as a restorative tonic. 'Taking caudle' became a custom for those visiting a new mother and baby, and 'caudle parties' became a fashion in the UK in Georgian times. So strengthening was the hot, honeyed brew that in France it had begun to appear earlier on the scene, and was taken to the groom in the middle of his wedding-night.

There was a time when the first blessing a midwife or a priest would give to a newborn infant was a small thumbnail of honey. Again, this was a strengthening tonic as well as a joyful welcome to the world. In India, it might be dropped into the mouth of the newborn from a golden ring or spoon. Today, however, beekeepers are urged to discourage this practice due to the limited risk of botulism.

Some of the great orators, prophets and kings throughout history are associated with legendary stories of bees arriving at their birth or soon after, sometimes alighting on their lips or building honeycomb inside their mouths while sleeping. In all cases this is an affirmation of a special birth, bestowing the child with an eloquent 'sweet speech' and a divine knowledge or wisdom. In one initiatory tale from the 'Ambrosian Life' (*Vita Ambrosiana*), when the ancient Greek poet Pindar fell asleep as a boy, bees gathered upon his lips and built a honeycomb there. Other versions of the tale tell us that bees settled upon his myrtle crib when he was a baby. All agree that he had close relationships with the Gods and the Muses. He wrote much about bees in his poetry, and he describes the gift of song as the 'Muses' Honey'.

Another apian prophecy occurred when a swarm of bees surrounded a newborn in the Ethiopian town of Roha in 1162. The baby boy miraculously emerged unscathed amid the town that his mother, a royal servant maid named Kirwerna, sheltered in to give birth far from the child's angry father, King Jan Seyoum. She saw the presence of the bees as a sign of her child's future power and gave him the name Lalibela, which means 'the bees recognise his sovereignty'. The boy became a man and eventually the bees' prophecy was fulfilled when he was crowned King Gebre Meskel Lalibela of the Zagwe Dynasty of Ethiopia and Eritrea in 1181. After a life of pious words and deeds he was made a saint by the Ethiopian Orthodox Church. The town Roha was renamed Lalibela and is famous for monolithic rock-hewn churches that were commissioned by the king.

Nearby are older, sixth-century, semi-monolithic churches that have housed many swarms and colonies of bees over countless years, like bee caves. The honey gathered there is known as *mar*, considered 'holy honey', and subsequently used in Christian healing ceremonies. There are three traditional ways it is administered by a priest: directly onto the skin or afflicted body part, by eating it, or by mixing it into holy water to be drunk. Accompanied by prayers and songs, this honey has been the salvation of many suffering Ethiopians, and a boon to the tourist industry during times of peace.

'Logos hive' – bees as the word, housed in an api-sculptural log-hive made by the author.

A similar story is known in the region, of bees arriving at the birth of King Solomon, affirming him as the rightful ruler-to-be of Israel and Judah, and further empowering the reverence of the honey churches and a faith in the power of the divine honey. The famed mead of Ethiopia, known as Tej, is also used as a holy communion.

Initiation rites

Many initiation traditions have included honey. Perhaps one of the most immersive of these is a ritual known as 'honey eating' performed in the Andaman Islands of the northeastern Indian Ocean. The shoulder of a young initiate is smeared with honey as they begin a honey fast. After then eating honey, it is poured onto their head, and they are then rubbed all over with it. This is followed by a bath of water, so they don't end up covered in ants.

Medicinal bees

The only consistent thing about bees is their inconsistency.

DR C. C. MILLER, 1831–1920.

This quote from the American beekeeper Charles Miller encapsulates the success of bees and the efficacy of their medicines. Inconsistency can be considered as one of their winning formulas. If a creature is predictable, it may be more prone to predation than those who are not. With well over 120 million years to hone their ways, bees have become experts at adaptation. They have also become highly skilled in collecting and improving some of the most effective antibiotic, anti-viral, anti-microbial and life affirming substances from their surroundings. Akin to medical herbalists and chemists, they gather these powerful ingredients to mature, ferment, ripen, convert and enrich. Inside their colony, eggs and larvae develop in the waxen hexagonal cells open to the air for the first eight days, until they pupate and are capped by the worker bees. To develop well and grow strong the vulnerable brood must be in an incredibly healthy environment. It may be a morbid thought, but one can better understand the power of this situation by imagining a pregnant woman with an open belly. With her foetus developing in a similarly exposed environment, the mother and child would have to be in a safe place full of salubrious substances. Bees have perfected such a place, shrouded in apparent olfactory darkness.

Everything that comes out of a beehive is medicine. The inconsistency of the chemical constituents found in the substances they collect also works well in their favour. There is a rich variety of nectars and pollen stored in a hive. A beekeeper knows that the sources of the nectar, pollen and propolis can vary from comb to comb, hive to hive, season to season and of course place to place. Along with its anti-microbial activity, this is one of the reasons why a honey wound pad can be more effective than a mineral-based pad. Although it will have to be changed more frequently, the healing process will often be more successful. Add to that the hygroscopic nature of honey, which essentially means it draws moisture into itself, including toxins and bacteria. Helping a honey massage, the treatment of a burn, an MRSA or one of the many other external or internal remedial ways in which honey has traditionally been used.

Honey and Longevity

Including good quality honey in one's diet will of course be of great benefit. In one Ayurvedic text, a diet is given that can prolong human life, of which milk and honey are the main elements. Pliny the Elder once stated that the ancient Britons lived to 120 years of age and put it down to their moderate consumption of mead.

The ancient Greek polymath and 'laughing philosopher' Democritus lived to a ripe old age because of his love of honey and mead. Hipparchus says he lived to 107. When asked about his longevity he declared, 'wet your insides with honey and your skin with oil'. The time came when he decided to die, and he simply gave up eating. However, the women of his household wished to celebrate the coming feast of Thesmophoria with him. After their entreaty he asked for some honey and prolonged his life by inhaling its aroma. After the feast it was taken away and he died.

Last rites

At the end of one's life, death and funerary customs are often graced with the golden liquid. Honey beverages and honey products may be the last nourishment a dying person may be able to consume, until one's last words and final expiration. In the wonderful 2019 documentary film 'Honeyland', made in Macedonia, the skep beekeeper Hatidže Muratova feeds honey to her dying mother. In parts of Africa, bee-brood (the uncapped eggs, larvae and pupae-stage bees in their cells) is traditionally the strongest medicine given to the oldest people in their weakest moments.

In ancient Greece and Rome, we see a tradition where libations of honey were made to the dead, the first of these being made at a funeral, and thereafter when the spirits of the deceased were consulted or placated. In more recent times, those wealthy enough were sometimes laid to rest in a coffin full of honey within family tombs. The bodies of some nobles were preserved in honey for voyages home across seas, and honey, wax and propolis were also used for embalming and mummification.

There is a curious legend in Asia where an elderly person around the age of 70 or 80 would give up their lives in service to their community and consume only honey for a month or so until death. They would then be sealed into a honey-filled stone coffin, and after a hundred years the resulting confection would be ready to be taken in small doses as a prized panacea known as 'melliferous man'.

Across Europe, the customary last rites of a beekeeper involve telling the bees of a death, often adorning the hives with a black cloth, and sometimes turning them to face a different direction. In France, a garment of clothing from the deceased would traditionally be buried in front of the hive. Considering the olfactory and pheromonal excellence of the creatures, this makes perfect pragmatic sense to me, being more than sentimental and sympathetic magic.

Don't you wait where the trees are,
When the lightnings play,
Nor don't you hate where Bees are,
Or else they'll pine away.
Pine away – dwine away –
Anything to leave you!
But if you never grieve your Bees,
Your Bees'll never grieve you.

RUDYARD KIPLING, 'THE BEE-BOY'S SONG',
PUCK OF POOK'S HILL, 1906

A jar of honey would be put near the newly departed as an important part of the traditional midnight repast in death chambers in Brittany. It was believed that a little fly would emerge from the mouth of the beloved dead and fly to the honey to take its fill. This fly is of course deemed to be the soul of the newly departed, preparing itself for the long journey to the spirit-world.

Putting pots of honey next to corpses was a common practice in many cultures, including Egyptians, Hindus, Chinese, Romans, and Greeks. Hippocrates, the Greek 'father of medicine', lived to the same age as Democritus, 107. It is said that a swarm of bees alighted on the grave of Hippocrates, and produced an exceptional honey that was effective against thrush, *Candida albicans*. This led to many nursing mothers making mass pilgrimages to his tomb to collect the special honey for their babies.

Bees as souls

As psychopomps, bees can guide one's soul to the otherworld, 'making a beeline' for the afterlife. The old Germanic expression 'path of bees' expresses this idea: that the air is filled with the souls of the dead. Across other parts of Europe, dreaming of bees could foretell a death. A very widespread belief is that a passing or visiting bee, moth or small bird is the actual soul of a newly departed travelling to the spirit realm, perhaps stopping for a while as a final farewell. Bees have such a blessed air about them, it was thought that if one flew into the mouth or ear of a corpse, it would spring back to life!

Such attitudes towards 'bee-souls' are also affirmed in a charming and ubiquitous tale found throughout various parts of Britain, of two travellers making their way across a moor.

Having stopped for a rest by a stream, eaten their lunch and drunk a bit of cider, one of them falls asleep in the hot sunshine. His friend watches in awe as a bee

The Widow, a painting of 'telling the bees' by Charles Napier Hemy (1841–1917).

emerges from the open mouth of the sleeping fellow, and steadily crawls through the grasses towards the stream. Shocked and spellbound, he slowly puts down his cider a cautionary distance away. Caught in a reverie he looks on as the bee reaches the edge of the water. It crawls up a fallen branch and crosses over the trickling stream, reaching mossy pebbles and a rocky bank on the other side. Safely there it crawls consistently over the pebbles and up the edge of a large rock. Atop of the rock sits an old ram skull, lying on its side. The bee ambles up the ridges of the ram's horn, and along the gleaming white skull. Then it enters straight into one of the eye sockets, and circles around the inside disappearing from sight. After a short time the bee reappears, retracing its steps exactly, and crawls back into his friend's mouth. The sleeping lad begins to stir then wakes up, stretching and yawning. He immediately recounts a vivid dream he just had of wandering through a forest to a wide river, where he crosses over a bridge to some rolling hills and a mountain. His friend listens on in amazement as he tells of climbing up the mountain to a gleaming castle with a grand staircase, and how he went in through an elaborate door and wandered around the fine halls inside, then returning the way he came, back over the wide river, and through the forest and then woke up.

Seamus Murphy, *St Gobnait* (1951), Ballyvourney, County Cork.

Honeyed words

The theme of bees predicting the divine power of poets and sovereigns is lubricated with honey. Lips that are symbolically and literally anointed with it bring forth truth and eloquence. The Thriae of Ancient Greece, the aforementioned tutors of Aristaeus, when fed upon the golden substance, uttered true prophecies, but had no such powers without. These three sisters are depicted with female heads, and the torsos of bees.

There are many paradoxes amidst the allegorical sweetness. If we say that someone has a 'honeyed tongue', this might warn of the opposite of truth-telling, insinuating a sweet skullduggery or a manipulative charm. When a person is described as mellifluous, however, we are not judging their veracity or trustworthiness, but suggesting a golden warmth and a soothing pleasure.

Bees, honey and the spoken word are intertwined in human heritage. The word for word and bee is the very same in ancient Sumerian – *Dabah* – from which the Hebrew name Devorah evolved, giving us Deborah today, which does indeed mean honey bee. The medieval Irish Saint Gobnait is sometimes called Saint Deborah, and she is one of the patron saints of beekeeping, who gained a feisty reputation after turning her bees against some marauders who were ransacking her abbey and village.

Saints and preachers, poets and orators, muses and gods, bards and oral traditions and their wise words soon reached the vellum and parchment of the holy books of the more recent doctrinal religions. The logos and the scribes were eventually collected into sacred texts like the Torah, the Bible, the Koran and the Ayurveda, to name just a few.

Blessed bee

The honey-bee paradox apparent in their ability to bring pleasure and pain continued into the Christian mythology with the intertwined suffering and redemption of Christ, reminding us of the idiom 'a bee can make honey from the bitterest of flowers'. In the Catholic tradition, the honey bee often represents the Virgin Mary, also known as the Queen of the Bees. In the Old Testament, the Promised Land is described as flowing with milk and honey, indicating abundance and fertility. The bee is a symbol of wisdom, for as this tiny insect collects pollen from the flowers, so humans may extract wisdom from the experiences of daily life.

Oshun of Africa and Afro-America is a goddess of love, joy, fertility and flowing fresh waters. Her colour is yellow, she is customarily offered honey, and used honey as a seductive charm. The ancient Egyptians considered bees to be a sacred gift from their solar deity Ra. As his tears fell from the sky and touched the earth they turned into bees, bringing honey which they called 'liquid gold', from the 'tears of Ra, or Re'. This was echoed in Christian mythology where tales tell us that the tears shed by Christ upon the cross turned to bees, or the water that fell from him after bathing in the Jordan did the same.

Bees and royalty

So highly did the Frankish ruler Charlemagne regard honey bees that instead of bestowing medals upon his commanders and officials, he would give them a beehive. He encouraged beekeeping throughout the realm, ordering that all estates had beehives, with two thirds of the honey produced given to the Crown.

Catherine the Great also valued bees and beekeeping, passing laws that protected hives and habitat, eventually representing a beehive upon her coat of arms.

Zosim

One of my favourite bee deities is the joyful and onomatopoeic Slavic god Zosim, a bringer of agriculture and beekeeping whose image was placed near hives for their protection. This deity was later Christianised as Saint Sossima, a patron of beekeepers. He is often depicted beekeeping alongside Saint Savvatiy, who may be a vestige of the pagan god Sabazius, and also has a buzz in his name.

Bees and chastity

The queen bee is the sole female within the colony who can successfully mate and lay a fertilised egg. Historically humans assumed chastity within the hive colony, as there seemed to be no evidence of sexual congress. It was thought that bees reproduced themselves miraculously. 'They neither rejoice in bodily union, nor waste themselves in loves languors', states Virgil in his poetic Georgics (Book IV), dedicated to apiculture and bee lore. This belief is one of the reasons why the symbolic beehive and its inhabitants were inspiring for religious movements teaching chastity. Throughout history, puritanical systems, monotheistic and polytheistic traditions, civilised states and cultures alike have viewed the beehive as completely chaste. Even when we discovered that bees did have sex, we were astonished that the mother could remain in a state of pregnancy for months on end, continually producing brood after brood. This was before we knew that queens would mate with up to 20 drones on their brief, but promiscuous nuptial flights, return to their colony and begin to lay eggs in the carefully prepared hexagonal cells.

The bears and the bees

Medved is one of the Slavic words for bear; *Med* being honey, and *Ved* being knowledge or wisdom. Thus, the bear is the 'honey-wise', the one who knows where the honey is. The children's tales of Pooh Bear parody this one-track mind for Hunny. Tenacious bears looking for bee colonies were a challenge for tree beekeepers, and indeed any early community and culture that co-existed with them under the sun, moon, and stars. The saying 'talk of the devil' is thought to have its origins in not mentioning bears, because people believed if you did utter the forbidden name, one would appear, and ursine chaos might break loose. The very name 'bear', meaning 'brown one', is an alias of their true name… which of course, I'm not going to mention.

The Bear and the Bee-Hives **by Samuel Howitt (1756–1822).**

ROYAL FAMILY BY LINEAL DESCENT.
QUEEN
LORDS
THE PILLAR OF THE STATE
COMMONS
THE BRITISH CONSTITUTION.
TRIAL by JURY
FREEDOM TO ALL RELIGIOUS DENOMINATIONS
LAW & EQUITY.
FREE PRESS.
HONEST and INDEPENDENT
MEDICAL SCIENCE
SCHOOLS
LITERATURE
ART
COLLEGES
CHEMISTRY.
INVENTION
WORK for MEN and BOYS
AGRICULTURE and FREE TRADE
GIRLS and WOMEN
MECHANICS.
BUTCHER
BAKER
GREENGROCER
CHEESEMONGER
TEA DEALER,
DAIRY
GLAZIER & JEWELLER
DRAPER
BOOKSELLER
TAILORS
BOOTMAKERS
HATTERS
WEAVERS
CARPENTERS
BRICKLAYERS
SMITHS
MASONS
UPHOLSTERERS
ENGINEERS
CABMAN
OSTLER
PAVIOR
SHOEBLACK
BOATMAN &c
COALHEAVER
SWEEP.
COSTER MONGr
DUSTMAN &c
CONDUCTOR
ROYAL NAVY
MERCANTILE
MARINE
BANK
OF THE RICHEST COUNTRY IN THE WORLD
ROYAL NAVY
MERCANTILE
MARINE
THE ARMY
THE NAVY AND THE
VOLUNTEERS.
Designed in the Year 1840 – by George Cruikshank and altered & etched by him in Febr 1867 – & Pubd in March/67
THE BRITISH BEE HIVE.
Published by the Artist and Sold by W. Tweedie 337 Strand – LONDON –

Chapter 2

Bees and Hive Symbolism

Symbols speak to the soul. They speak without words, directly to one's heart and subconscious, in the realms of the *mundus imaginalis*. They are transrational and can signify different things to different people. A bustling skep may suggest an organic city, or institution, healthy and well warded. It could be a thriving community, a dedicated family, or an inspired individual. A skep full of bees is a protected place full to the brim with a powerful sense of purpose and prosperity. The symbol of the beehive, originally in its skep form, has been utilised in art and architecture, religion and spirituality, business and commerce. It has been portrayed to convey various meanings and messages – through heraldry, painting, politics, commerce, banks, public houses and business – in art, religion and spirituality.

Left: **George Cruikshank, 'The British Beehive', 1867 (see page 42).**

Below: **A golden hive over the door of Lloyds Bank, Fleet Street, London, UK.**

The UK

The phrase and engraving on page 42 features on the title page of the *Summum Bonam*, by the British physician and alchemist Dr Robert Fludd (1574–1637). The Latin translates as 'The rose gives the bees honey'. It symbolises several things, including the journey of the soul through the pilgrimage of life. The rose is a labyrinth, being a thorny but sweet destination. Once hardships and honey have been gathered, the pilgrim returns to the hive with great benefit to their

***'Dat Rosa Mel Apibus'*, – 'The rose gives the bees honey' – from the frontispiece of Robert Fludd's 1629 *Summum Bonum*.**

community. Beekeepers and botanists may roll their eyes a little bit here if they know that honey bees collect pollen and not nectar from roses, but the allegory is still a strong one, and not to be taken literally.

The beehive was once a common sign under which any wayfarer or pilgrim might enjoy a well-earned drink. Many alehouses, taverns and inns were named after this symbol of community, hospitality, and health. It is possible that these were meeting places of trade guilds, fraternal orders, and friendly societies, like the numerous Oddfellows Societies who adopted the bee skep as one of their symbols. At a time of widespread illiteracy, these signs and symbols were key. The ale that I grew up drinking, Brakspear & Sons, sports a bee in the logo.

The hive and social order

In Britain in 1840 the book illustrator and caricaturist George Cruikshank (1792–1878) designed 'The British Beehive' (see image on page 40) as an anti-reformist comment in support of not expanding the franchise to the public voter. He didn't etch and print it until 1867 when the Second Reform Act was going through parliament. The cartoon depicts the hierarchy of the British workforce housed inside a giant bee skep. It echoes the ideas of the Scottish philosopher and astronomer, Adam Smith. In the 1770s, Smith published and taught his theories

Napoleon and the bee

Today we know that a beehive is a self-directed collective with no autocratic manager nor benevolent dictator. However, societies modelled upon beehives were viewed quite differently in the past. Pharaohs, popes, monarchs and rulers and many in positions of power justified their autonomy with bee symbolism. Napoleon Bonaparte is perhaps the most famous ruler to have adopted the bee as an emblem, linking him to earlier dynasties. His Napoleonic Code used bee symbolism to express hard work, order, loyalty, and commitment to the empire as 'Queen Bee'. This sentiment has since been broadened in contemporary spiritual traditions, fostering a commitment to the queen bee as mother nature herself.

that social status and social identity were primarily determined not by social, religious, or political rank, but by occupation and by an individual's relation to the means of production. He is often quoted for his metaphor of the 'invisible hand', which he used to describe unintended social benefits resulting from individual actions, suggesting that the successes of the wealthy naturally trickle down to the poor through the development of the nation. Cruikshank, Smith and many others wanted to preserve the status quo. As we well know, subsequent acts were passed and the right to vote was continually extended to more people, as equality and reform proliferated.

Crowned at the top of the cartoon of the taxonomy of trade and craft is Queen Victoria flanked by the royal family and underpinned by the British Constitution of the Lords and the Commons. Down another layer is the priesthood, entitled 'freedom to all religious denominations' and flanked by Jury and Law. 'Agriculture' holds a large place at the centre, topped by the 'Free Press'. Then represented throughout are the butcher, baker, greengrocer and so forth. Right at the bottom of the 54 'cells' representing the various trades and classes of the British workforce and society is placed the military – the foundation of Cruikshank's representation of British society. To a beekeeper or entomologist this makes perfect sense. They are the guard bees at the 'city gates', checking every entrant for hive scent and good intention. There they stand, a venom-armed guard, who in a natural state would be warding an entrance enhanced or 'curtained' with the miracle substance propolis.

The USA

Many Europeans looking for new starts in life and religious freedom adopted skep beehives as a symbol of swarming to the milk-and-honey lands of opportunity. Utah was named the 'Beehive State' after the Mormon settlers' adoration of the honey bee, or 'Deseret' as they knew her (and had wished to name their state). Their whole religion was inspired by a dream of golden plates (perhaps symbolic of honeycombs?) with hieroglyphics upon them that, when interpreted, became the Book of Mormon. Many New World colonialist American dreamers were employing the utopian beehive, which became symbolic of the new beginnings, order, and prosperity they were seeking. As Tammy Horne states in her book *Bees in America*, they considered themselves to be 'swarming' or 'hiving off' to pastures new. Colonies of honey bees travelled across the Atlantic to the Americas in skeps and 'bee gums' (a term for log-hives at that time). I imagine the best survival rates

An 1850 emblem for Utah, 'the beehive state'.

were in those colonies taken over during the winter months, whilst nestling in with the stores. The bees travelled further and faster than the human colonisers that took them to the Americas. They became known as the 'white man's fly' to the indigenous occupants, and often preceded the colonisers' presence.

Beehive buildings

Many architects and town planners over the years have shared similar inspiration from bees and bee lore. From the dry stone 'beehive' huts in Ireland, believed to have been built by Celtic monks, to the New Zealand 'beehive' parliamentary building. The latter was originally designed by the Scottish architect Basil Spence, with the Sussex University campus and its 'multi-faith chapel' among other examples of his hive-minded concepts.

Bees, as winged creatures, have also inspired airports. It was Morris Jackaman who conceived the idea for the central 'Beehive' building of the original Gatwick Airport, giving it a benevolent apian theme. Heathrow airport also sported a strong hexagonal design in the layout of the original runways, being pragmatic for the early aviation era when planes had to take off facing the wind, and akin to a protective 'Seal of Solomon'. Joseph Beuys, Antoni Gaudí, Mies van de Rohe, Le Corbusier, Wolfgang Buttress, and recently Gianluca Santosuosso are just a few notable names on the long list of bee-inspired artists and architects, all sensing and celebrating bees and nature as creative strategies for our lives and lifestyles – some of them becoming champions of the plight of bees themselves.

Top: **The 'Beehive' at Gatwick Airport, London, UK (opened 1936).**

Above: **New Zealand's 'beehive' parliamentary building, inaugurated in 1977.**

Symbols of industry

The scene of a straw skep on a wooden stand, in a verdant and blossoming setting, with golden honey bees diligently issuing forth amidst the illuminating rays of the sun, has inspired many societies and institutions with its symbolism and allegory.

So work the honey-bees,
Creatures that by a rule in nature teach
The art of order to a peopled kingdom.

William Shakespeare, *Henry V*
(Act I, Scene II), 1599

Above: **The Apis Club coin, 1919.**

This iconic, heraldic image of the skep hive appears within ancient mystery schools, Alchemy, Rosicrucianism, Freemasonry, Druidry, Oddfellows societies, the Daughters of Rebekah, the Knights of the Maccabees, the Order of the Bee, the Catholic Church, the Church of Jesus Christ of Latter-Day Saints (Mormons) and the State of Utah, Lloyds Bank, the Co-operative Group, and many other businesses and social groups.

> The beehive is found in Masonry as a reminder that in diligence and labour for a common good true happiness and prosperity are found.
>
> MANLY P. HALL,
> *THE SECRET TEACHINGS OF ALL AGES*, 1928

Above: **The Beehive public house in Brentford is a magnificent building featuring a skep turret and skep signs and has been recognised as having sufficient quality to be given protected status and saved from development. Brentford's famous football team's nickname is of course the Bees.**

Above & right: **A silk-embroidered multi-degree apron (and detail of the bee skep depicted on it). Worn by Freemasons during rituals, the apron contains allegorical symbols that teach a peculiar morality.**

SOLERTIA ET LABORE.
SOLERTIA ET LABORE.

Chapter 3

Revd Charles Butler

The 'Father of English Beekeeping'

As of all states the Monarchie is best,
So of all Monarchies that Feminine,
Of famous Amazons excels the rest,
That on this earthie Sphaere haue euer bin

REVD CHARLES BUTLER, 'MELISSOMELOS',
THE FEMININE MONARCHIE, 1623

The second edition of *The Feminine Monarchie*, published in 1623, was made bigger and better than the first edition of 1609, for which Charles Butler hadn't expected a large audience. Most marvellously this new publication had the 'Melissomelos' or 'Bees' Madrigal' set right at its heart, as the book was designed to be laid or held open and the piece sung by four choristers who would gather around it. The Bassus and Contratenor parts were printed upside down on the upper half of the pages, the Mean and the Tenor parts the right way up on the lower half. This enabled everyone to read their part with ease, facing each other, which must have made renditions a more intimate ritual. The melodies and harmonies are inspired by the composer's experience and knowledge of listening to bees, their murmurings, refrains, chants, and moods. At the height of the piece, all stills as the performers begin to intone and mimic the piping of virgin queens. Throughout the book, one of the most ancient appellations for bees, 'the muses' birds', is employed by Butler, insightfully reassuring us 'that if Musicke were lost, it might be found with the Muses Birds'.

Depiction of the Revd Charles Butler, St Lawrence Church, Wootton St Lawrence, Berkshire, UK.

Charles Butler was directly quoting a much more ancient poet, Virgil, in his reference to 'Muses Birds'. His great love of antiquity shines throughout his books. Perhaps his knowledge of the classics, of the older pagan traditions that exalted divine femininity and worshipped goddesses, instilled within him a deeper respect for the feminine and for nature? Declaring that bees have been made by God for the use and service of humans, and that they are evident proof of the infinite power and wisdom of the creator, shows us he was a faithful devotee of his Christian mythology and doctrine. He was, however, an eccentric and open-minded member of the clergy, to say the least. He didn't use the Christian calendar to calibrate the year, preferring the western zodiac as a more accurate timekeeper, and advising the reader that in Pisces it is safe to remove a mouse guard, in Virgo to make a first harvest, there is forage until Sagittarius, and suchlike.

> The Melissean yeare is most fitly measured by Astronomicall months … begin with the Sunnes entrance into severall signes of the Zodiack, and are therefore called by their names … the Sunne, entering into twelve signes … so beginning the twelve moneths, doth notoriously alter his course making the days longer or shorter the aire warmer or colder the earth more fruitfull or barren making also both of the Aequinoctia and Solstitia.
>
> CHARLES BUTLER, *THE FEMININE MONARCHIE*, 1623

Summer door and winter door.

As well as the classical pre-Christian writers strewn throughout the text, Butler adheres to the traditional star-lore, and the 12 metering signs of the Zodiac as ever-present and pragmatic timekeepers. Of the bees, he writes: 'As skillful astronomers, they have fore-knowledge of the weather'.

Like a conductor, he gives the beekeeper the rhythm of the Melissean year. He wrote also of the asterism the Seven Sisters or Pleiades nymphs: that heavenly flock of birds, starry swarm of fireflies, and doves of the muses clustered within the neck of Taurus. Butler quotes Virgil, stating that their sighting and position fixed the two seasons of the honey harvest, along with calculations from the Equinoxes.

Butler is eccentric and entertaining to say the least: metering the year by the movement of stars, constellations, and asterisms; using the names of ancient Amazonian queens to define primary, secondary, and tertiary swarms; composing an onomatopoeic madrigal; delighting in ancient pagan mythology and philosophy; perpetuating folk customs and supernatural charm; correcting regurgitated misconceptions with practical knowledge; and imparting wisdom and medicine. Writing later, in the autumn of his life, Butler published a book solely dedicated to song and music theory: *The Principles of Musik in Singing and Setting.* He exercised a healthy disruptive streak. As a language reformist he published a book condemning traditional English spelling in favour of a phonetic alphabet, and as a Protestant and parent he wrote a theological defence of marriage between first cousins.

> At Gemini set the doores wide open
> At Virgo… set up the Winter-doores,
> and fasten them with cloome
>
> CHARLES BUTLER, *THE FEMININE MONARCHIE*, 1623

Below: **Winter door in situ.**

Below right: **Mouse guard.**

> [T]he philosopher ... he every where calleth the governor Rex ... as many as followed him, searching no further than he did, were content to say as he said .... So that I am inforced to translate it Queene ... this being an Amazonian or feminine kingdome.
>
> CHARLES BUTLER, WOOTTON, 30 MAY 1623

Luis Mendez de Torres wrote in his *Treatise on the Cultivation of Beehives*, published in 1586 in Spain, that the large bee at the centre was not a master and king, but a mother and queen. He called her *maesa*, a female version of *maeso*, master – as well as the *maessa de enjambre*, which some translate as 'mistress of the swarm'. Bees and flies were also, however, commonly identified in witch-trials and 'daemonologies' as familiars, spirit-helpers, or shape-shifted practitioners. In the same year that Charles Butler published his first book, *The Logic of Ramus* (1597), James VI published his handbook of the supernatural world, *Daemonologie*, endorsing the practice of witch-hunting.

Detail of Butler's memorial window showing three skeps under a shelter.

Author introducing bee skeps and the 'Melissomelos' madrigal to Queen Camilla, President of the charity Bees for Development.

Throughout Butler's developmental years at Oxford, powerful feminine monarchs were vying for the throne. A contemporary of Shakespeare, whose defiant female characters speak to a proto-feminist current alive at the time, Charles Butler reissued *The Feminine Monarchie* in 1635, dedicating it to the Queen Consort, Henrietta Maria of France, a strong-willed woman well able to exercise her power and influence. Butler, unwittingly perhaps, was dropping a small yet magnificent piece of proto-feminist medicine into a world that was particularly rife with patriarchal pressures escalating the oppression of women. All this rich history and heritage are stitched into the humble bee skep.

The Feminine Monarchie is a comprehensive celebration of bees, rooted in the ancient texts and mythologies, yet expanded with much practical advice from an author who wasn't simply regurgitating what the ancient masters had penned, but was also questioning, challenging, and correcting them. A clear royalist and Protestant ideology is visible. It is unlikely we'll ever know if there was any intended gender politics and feminist allegory in the work, but he certainly praises Amazonian rule, and the voice of the queen calling forth secondary swarms: 'the females have the preeminence … the feminine gender is more worthy than the masculine'.

On the almost egg-shaped form of the skeps he used, he wrote:

> The bees do best defend themselves from cold, when they hang round together in a manner of a Sphere or Globe (which the Philosophers account the most perfect figure) and therefore the neerer the Hive commeth to the fashion thereof, the warmer and safer the bees.
>
> Charles Butler, *The Feminine Monarchie*, 1623

Within the 'Mellisomelos' – the Bees' Madrigal, the Queen of the Amazons, Antiope, is lamented as the prime swarm departs, heralding the ascent of her sister Queen Orythia to the throne. Once a year on 20 May, now ingrained into the beekeeping annual, we have World Bee Day, the birthday of Anton Janša, who holds the title 'Father of European Beekeeping'. The question might also be, who are the mothers of beekeeping?

Chapter 4

Making a Skep

A step-by-step guide on how to make a bee skep

As mentioned earlier, a 'skep' is simply a basket, and 'bee skeps' are baskets for bees. They have survived in the UK, Europe and the USA, primarily as swarm-collection vessels. In recent years, there has been a slow resurgence in the craft of skep beekeeping. Compressed straws are good insulation, and the bee skep's shape – usually rounded – suits the bees' natural cluster. Form follows function, and the archetypal bee-skep shape is a kind of bell-like, almost parabolic dome. The structure is formed by one continuous spiral of coiled straw from the top to the bottom. Each circuit of the straw, sometimes known as a 'rowl' or a 'lissom', rests upon the one below, transferring any weight down through the walls to the floor. One can happily stand, and if inclined bounce up and down, on top of a well-made skep, which ought to outlast its maker if looked after.

When permanently positioned to receive a swarm, a bee skep of any kind is traditionally dressed in a composite material like cob (subsoil, sand, chopped straw, and water) or 'cloom' / 'cloome': a mixture of cow dung, or 'neat's dung', tempered with wood-ashes, lime, or sand. It isn't waterproof, but adds an extra layer of insulation and strength, and protection from the damp. The skep may be sealed to its floor using a similar substance. These floors, or 'bee boards', were and are more commonly made of wood, and less commonly of stone in areas where that is more prevalent. Stone is less preferable as it becomes very cold in the winter, and very hot in the summer. The finest floors are chamfered around the edges to shed any water. Some bee skeps have entrances, and others don't. In the latter case, if a skep has no entrance, one can be cut into the floor as a sloping channel or rebate. I'll write more about floors later. If you are making a skep solely for swarm-collection it doesn't necessarily need dressing, doors, or floors.

Charles Butler-style hive, stitched up with split-willow lapping, made for St Lawrence Church, Wootton St Lawrence.

Size

The size and capacity of a skep can also vary wildly. Their volume was traditionally measured in pecks, a peck being equal to nine litres. One of the few diagrams in *The Feminine Monarchie* shows the dimensions and form for what Butler calls a 'middling' or medium-sized hive. It measures 43 cm (17 in) high, 38 cm (15 in) at its middle width, and tapers in again to 33 cm (13 in) at the bottom rim, making it slightly egg-shaped, but broad enough to be stable. As mentioned earlier, Butler agreed with philosophers who exalted the sphere as the most perfect form and believed that the closer a skep could be made to a sphere, the better.

Most other bee-masters up until the late nineteenth century kept their bees in skeps that were the classical dome or bell shape. In more recent times they were made with flatter tops for double-hiving; there will be more about that in the coming chapters.

Skep-making in the sunshine. It is always nice to work outdoors when the weather allows.

Straw

Well, let's get started. As the saying goes, 'it's not rocket science', but there are of course a few tricks to the trade and nerdy nuances to negotiate. Whether you paddle your way forward like ducks to water, or plod along slowly towards self-assured success, you'll do well to start with good quality materials.

A 'Rolls-Royce' of straws is a good long straw wheat sourced from a thatcher, or a long-stemmed biodynamic rye from a farmer. One may find many varieties of wheat used for thatching, from Squareheads Master, Maris Widgeon and Aquila to Triticale and N59, some being sturdier, others softer. All are of good length and longevity and perfect for lipwork skep making.

However, it is important to know that one cannot make a skep using the common thatching material often known in the profession as Norfolk Reed, being the common reed *Phragmites*. This plant is much stiffer; the clue is in its Latin name. Being brittle, it will break and fragment if one tries to coil it. Most roofs that are thatched with Norfolk Reed are still ridged with long stemmed wheat straw. I've only ever heard of one thatcher who could ridge a roof with reed, an Irishman named 'Kneecap McGraw'. Apparently, he never told anyone else his reed-ridging trade secret. Perhaps he was soaking them to make them pliable enough, or using them green, or had a clever way of placing them? Anyway, I hope the story will help you remember not to purchase common reeds for skep making.

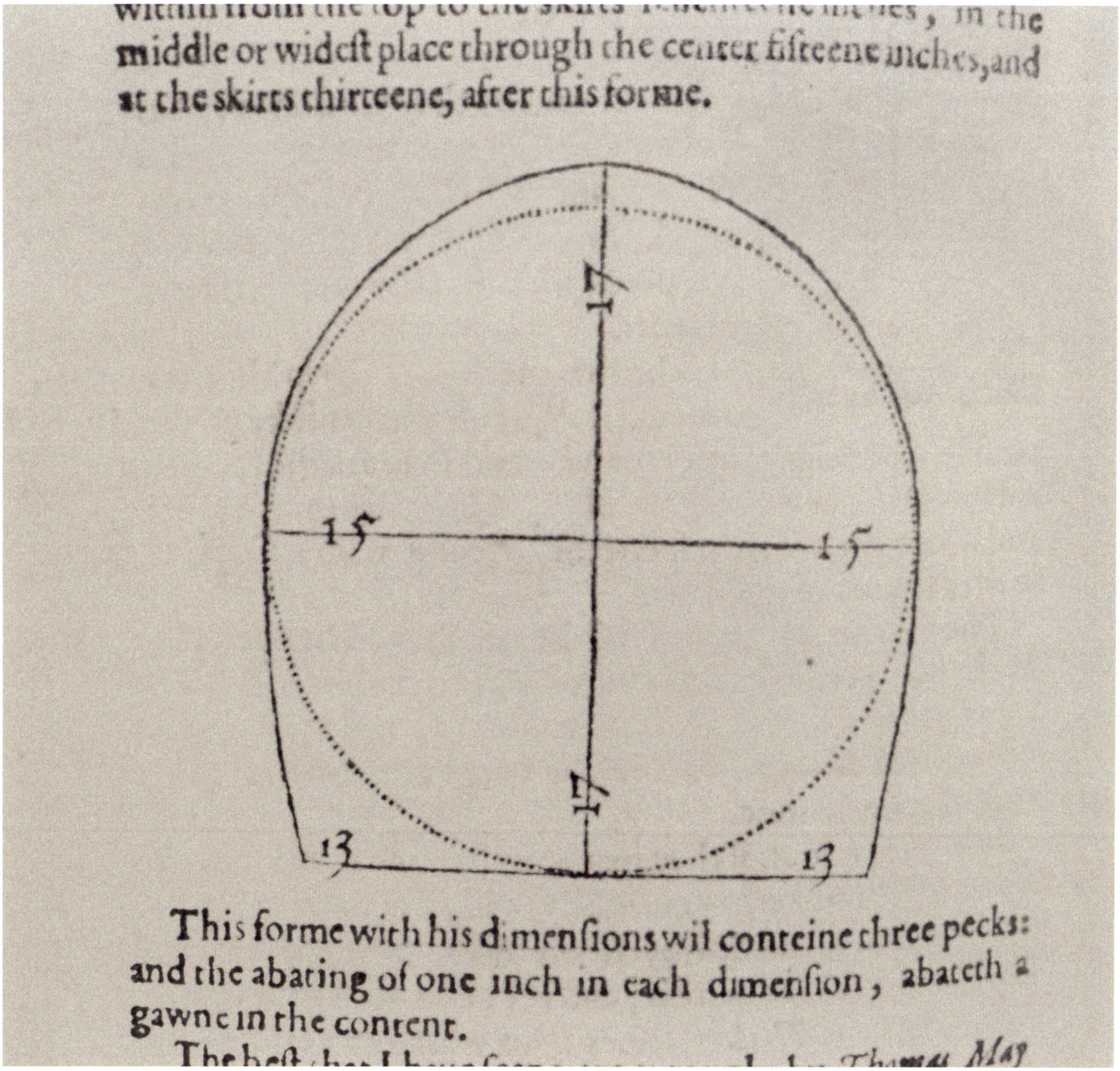

middle or wideſt place through the center fifteene inches, and at the skirts thirteene, after this forme.

This forme with his dimenſions wil conteine three pecks: and the abating of one inch in each dimenſion, abateth a gawne in the content.

Middling hive dimensions from Charles Butler, *The Feminine Monarchie,* 1623.

Most other sorts of grasses will do. In some parts these may be known as 'bennets'; up in Yorkshire the author Frank Alston employed purple moor grasses; down in the New Forest, Frank Vernon wrote about the 'Hampshire Pot', a regional variety of skep commonly made by gypsies from sedges and rushes. Across the world, we see many different fibres being twisted and bound into baskets.

Again, I do recommend starting with a long-stemmed wheat from a thatcher. There's an art to picking up the straws from a bundle without making an absolute mess, and it's unlikely any of us will learn that in a day. Often by the end of a skep making class, everyone is treading on or neglecting the perfectly useable straws they've dropped, to reach the neater ones in the bundle. Usually, however trodden or trampled on, it can all be used, and the debris can go on the hearth fire, the compost, to one's hens or even one's bee-smoker if your honey bee praxis involves that. The straw does benefit from some processing before one begins. It needs wetting, cleaning, and oftentimes massaging, especially at the beginning where it will be turning a tight spiral. A skep is essentially one continuous spiral of straw from top to bottom, although usually made upside down. Some skep makers like to completely saturate the straws with water, and then flatten them all on a bench-top with a mallet. Others like to spray the straw with water from a spray bottle. If one is working outdoors in the summertime, the straws will dry out very quickly, and it's best to keep your materials in the shade, and cover with a wet towel. Only soak as much straw as you need, as it will start to turn grey with mildew if stored away wet. One 'bundle' of thatching straw is enough material to make at least one large skep.

Lapping

So, what is lapping? It's a split material, most often used to bind another material: the straw of a lipwork skep; the branches of a birch besom; the rods in the handle of a wicker basket. Traditionally, split bramble and split willow were commonly used for binding bee skeps and other lipwork creations like seed baskets and the iconic 'Monmouth chair'.

Interestingly, honey bees benefit greatly from both bramble and willow. The bright yellow willow pollen in springtime is a perfect food for the strong build-up of bee-brood, and the bramble flowers later in the year are an abundant source of greyish-green pollen for the autumn brood, and nectar to ripen into honey stores in the winter months ahead. In my neck of the woods, these are the local, sustainable sources of lapping for lipwork. In other parts, lapping is known as lacing. One is never far from a bramble, and coppiced basket-willow produces excellent rods that can be cleaved into two, three, or four, depending on the diameter. I encourage you to get to know these plants as sources for cordage and bindings and more. However, for your very first skep, unless you are a competent craftsperson, experimental archaeologist, or a seasoned primitive technology nerd, I recommend using a 6 mm (0.24 in) lapping-cane sourced from a craft supplier or a seat-weaving supplier. This may come natural or glossy. Natural is a bit cheaper, but glossy can be marginally stronger. A 5 mm or 4 mm (0.2 in or 0.16 in) lapping-cane will also do. This is not the local, sustainable option, but it is very forgiving and beneficial for beginners. It has a uniform strength and a satisfying length, and all you need to gain confidence and competence with the lipwork process. Once you have learnt the craft and become confident with the techniques, if you're still keen, it is well worth delving into the richer, rewarding world of local hedgerow options. They are free and may well deepen that sense of seasonal connections to the landscapes we inhabit, strengthening a sense of belonging and wellbeing. I have found much

Repairing a lipwork 'Monmouth chair' with split bramble lapping.

succour through devoting time to indigenous crafts and traditions. They may be long and slow processes, and less alluring than screens and parties and the more instantly gratifying pursuits, but they are generally more nourishing, therapeutic, and strengthening overall.

There's a part of me that feels sad to recommend material with air miles, that has travelled a very long way to be here from places like the East Indies and has been milled by a machine from rattan cane. Hey ho, most of us drink tea and coffee, rather than roasting acorns and dandelion roots, and as I've suggested, once you're coiling up baskets in your sleep, jump onto whatever your local materials may be. For now, you have enough on your plate learning a new craft, and I'm trying to teach lipwork in the most accessible way. The lapping will really help you with your first steps. Approximately 200 to 250 grams will do for one skep.

Tools

The traditional working tools are a bone fid and a cow-horn feeder, and in useful addition to that a strong knife, secateurs, scissors, willow-cleave, and a bucket of water or a spray bottle. Waterproof leggings or an apron are advised if a wet lap is to be avoided. For the adventurous, a hardwood or horn-cleave for splitting willow or bramble into two, three or four, and a small wooden bat with varying sized holes drilled into it for running thorns off the bramble vines. Other top tips for thorn removal are ring spanners, forks in trees, stiff garden gloves or holes drilled into the side of the cow horn feeder. Bone and horn are both durable and accessible materials from time immemorial; most commonly goose bone and cow horn. Since an outbreak of bovine spongiform encephalopathy (BSE) in the UK in the 1990s, cow horn is classed as a toxic material. So don't 'put that in your pipe and smoke it'. Large bird bones are very suitable for making lipwork fids and needles, as they don't need much hollowing out and they are of a suitable size. One may fashion bone needles from mammal bones, notably deer, goat and sheep, but they are thicker walled and need a bit more whittling down to size. It is a good idea to add a lanyard of some sort as one may often find oneself literally looking for 'a needle in a haystack'.

Today I teach lipwork students to use Swedish rope-splicing fids, sourced from marine suppliers – referred to throughout this chapter as your needle. I also make them feeder gauges by recycling 40 mm (1.57 in) plastic water pipes, cut approximately 40 mm (1.57 in) long.

Cow-horn feeder.

Rope-splicing fid and traditional bone fid.

Bone fids, bone needles, and a horn three-way cleave.

Splitting willow to make laps.

Wooden bat for de-thorning brambles.

Splitting bramble.

Making the bramble supple.

1. Making the first rowl

Each circuit of straw is called a 'rowl', or 'roll'.

1. Wet and prepare the straw by cleaning it and discarding the leaves and ears. Scissors are handy here, or strong teeth.
   Have a bucket or a trough of water handy and soak a couple of laps. (Look closely at your lapping-cane and see that it has a flat side and a curved side. If you are using bramble or willow, the flat side is the inside, and the curved side is the outer, green side. The flat side will be bound onto the straw.)

2. Select your first lap and snip a point into one end.
   Tie an overhand knot in the other un-snipped end, with its flat face on the inner concave part of the bend and the curved face on the outer convex part of the bend. Tighten up the knot until the hole in the middle is about the same width as the lapping, the un-snipped tail being proud by about 25 mm (1 in) (**A**).

3. Have some properly prepared straw within reach. If sitting, lay the knot end of the lap on your thigh, the pointed end trailing away from you with the flat surface facing upwards. If standing at a workbench, do the same on its surface.

4. Pick up a medium-sized handful of around 30 to 40 clean, wet straws and butt up the ends by tapping them on the floor or bench top.

5. Place them on top of the lapping, close to the knot, with the butt ends on top of the lap with few centimetres/an inch or two of clearance, and the straw pointing to the right – some left-handed people may want to work with the straws pointing to the left, but not all (**B**).

6. Finger and thumb your way along the lap to find the pointed end, without twisting it. Insert the point down through the hole in the knot and pull the rest of the lapping through until you have a secure bind on the straw, with the flat face of the lapping in contact with it. Be careful to avoid paper cuts from the lapping and the straw.

7. Tighten up the knot. Keep it on your side of the straw and trail the lapping underneath the straw and away from you again, with its flat face upwards, and its curved side down. Finger and thumb along it to find the pointed end again. Circle it up and back down towards you and through the knot (**C**). If the hole in the middle of the knot has tightened up too much, insert your skep-maker's needle into it and wiggle to open it up a bit.

8. Tighten this second bind onto the straw, to the right of the first bind with about 3 mm (0.1 in) between them, progressing dextrally, to the right, along the straw, away from the butt ends (to the left if you are working the left-handed way).

Keep everything damp by spraying with water or dipping it into your water container (**D**).

1. Making the first rowl (cont.)

9. Repeat this technique to make a third bind around the straw.

10. Holding up your work with the lapping hanging down, make a bend of approximately 120 degrees over the knot (**E**).

11. Keep on wrapping the straw, feeding the lap underneath it, then arcing up and back around towards you and down through the same hole in the knot (which might need opening with your needle).

    Make sure to keep the knot in the middle as you continue to bind and bend the straw around it with each new wrap. Keep the first wrap nice and tight, so it doesn't slip off. If it does, you can slip it back on with care and tighten it.

    The straw will keep compressing, so your original wraps will become loose and need tightening up. Get into the habit of bringing any slack forwards using your needle to help shunt any loose lapping around the straw. Be careful not to snap the lap.

    Keep wrapping and stitching the pointed end through the knot; this will become the middle of a circle. Try to place the lapping equidistantly on this first circuit, with gaps of about 3 mm (0.1 in) between laps around the circumference.

12. When you have gone full circle, reach for the butt-ends of the straws, and snip them off with a good pair of scissors or secateurs as close as you dare to the first bind (**F**).

    Keep stitching into the knot in the middle, opening it up by wiggling the needle into it if you need to. If you have an open hole in the centre of a few millimetres, that's fine. Keep going as you pass the snipped end of the straws, covering them with the straws that will become your second circuit. Once you have buried your first ever wrap, it's time to start binding this second circuit of the spiral to the first.

2. Making the second rowl

1. To start the second rowl, insert your needle through, catching half a dozen straws underneath the second or third lap on first rowl, as you have most likely buried the first by passing it. If you have a selection of needle sizes, use the smallest one for this manoeuvre, being careful not to snap any laps or break free any straws (**A**).

2. Wrap the straw with the lap and feed it down through the needle, which creates a channel through the straw and underneath the lap on the inner rowl. Pull both out and tighten up the stitch. Repeat with each consecutive lap.

   Try to position the laps on the straw so that they are pointing at the centre like the spokes of a wheel. To achieve this, you may have to bend the straw of the second rowl considerably. Be firm, the straw won't break: push it into place, position your lap correctly, then tighten it up.

3. This is a good time to put your feeder gauge onto your work. It will help hold the trailing straws neatly, and you will be able to feed new straws into it. You'll have to clamp all the straw together with one hand whilst sliding the feeder on with the other. Some skep makers like to start with a small diameter feeder on from the start, and then swap it over with a wider one as they get going. You may find this to be a help or a hindrance.

   You'll need to regularly introduce new straws into the feeder from now on, each time you make a new wrap and stitch.

   To add more straws, move the feeder up close to where you are joining the rowls. Take a half a dozen straws at a time and feed them in butt end first. Make sure the ends of the straws are buried in the middle of the bundle, so they don't stick out when you turn the corner (**B**).

2. Making the second rowl (cont.)

4. Eventually you'll be working with the feeder at full capacity. From now on keep it full by introducing more straws whenever you need to. You'll have to do so more often than you think!

5. Every time you wrap the straw with the lapping, make sure to insert the needle underneath a lap on the inner rowl, diagonally, at around 45 degrees. Be careful not to push the needle too brashly, but slowly wiggle it through, to avoid puncture injuries, or stigmata (**C**).

   Keep wetting the straws and lapping, bringing any slack in your stitches to the front. If you can wiggle a lap, it can be tightened. The tighter the straw is bound, the stronger the skep will be. Be careful not to snap the lapping.

   When tightening up a wrap, give the straw a slight twist in the same direction as the lapping is running. This will give the whole form nicer curves. Try not to over twist the straws or hold on too tightly or it may become ragged. Remember that everything goes better when dampened with water, from a spray bottle or a wet sponge.

   You can start to relax into the work now, putting more effort in when tightening up the laps and compressing the rowls, relaxing when inserting the needle and feeding through the laps.

3. Continuing down the skep

Soon you will need to engage what I call the 'thumb-gap rule'. If the gap between the laps on the rowl that you are binding onto is wider than your thumb, then insert the needle into that gap. Catch about half a dozen straws and insert the lap through the needle as if you were stitching under a lap as usual. Tighten up. Then make sure you next insert the needle under the next lap on the inner rowl and continue stitching under the laps until you reach another gap wider than your thumb. This prevents your wraps from becoming too widely spaced, maintains strength and a regular distance between the laps, keeping a finer shape (**A**).

A

B

4. Joining in new laps

At some point you will come to the end of your lap. Keep going and keep tightening, until you can't make another stitch. Then go back and tighten up all the wraps you have already made. Using the needle to help you, try to bring any slack found to the leading end of the lap. If you can wiggle the laps, then they can be tightened. Be careful not to snap the lap, using the back of the needle. You'll develop your own technique, and your wrists will strengthen through the practice. I use my dominant hand to clamp a lap between the needle and my thumb, to then shunt it forwards, whilst compressing the previous lap and straw with my other hand. Keep damping it down to enable it to compress satisfyingly. Once you've done a good job of tightening, and brought any available slack forwards, you'll be able to make one or two more stitches. It is important to do this before introducing a new lap.

There are a few different ways to add a new piece of lapping, but I usually use the half-hitch or the blind join.

Half-hitch

Prepare a length by wetting it and snipping points into both ends. Put the needle in (or leave it in) underneath the last stitch you have made. Leave the head of the lap that has just run out free. Insert the new lap into the needle, with its flat face looking up at you, underneath the curved face of the old lap (**A**, **B**).

Make a foothold on the new lap by looping it under a lap on the inner roll, on the inside of the basket. You are looking at the inside ceiling, essentially making the bee skep upside down. Tighten up the foothold as a half-hitch (**C**). The free head of the old lap may now be neatly overlapped by the new one as you continue to make stitches around the straw as before (**D**). This ensures a continuous strength as the tightening of the new lap binds the old one and brings it slightly forwards. 'Free your mind and keep your feet on the ground', can be a mnemonic device to help you remember this way of introducing a new lap.

C

Blind join

I call this a 'blind join' because it becomes neatly hidden by the succeeding rowl. It is satisfying if done well, and who knows, one may even accrue extra marks in a Honey Show skep class with this technique.

Take the foot end of the new lap and insert it flat side up underneath two of the previous wraps of the lap on the outer rim of the rowl, perpendicular to them, parallel with the straws. Place it in where you think it will be covered by the next rowl when it circuits around later to cover the rowl you are currently creating. Then bring the head end of the old lap down on top of the new lap, flat to flat, crossing at right angles. Pinching the two together where they cross, make a twist so that they change places, the new lap now running across the straw with its curved side up, and the old head parallel with the straw with its flat side up. Keep a thumb or finger on the join whilst you make a few wraps with the new lap (**E**).

D

E

5. Working the dome

You're now well on your way to making a bee skep.

Remember to keep feeding your feeder, working with it at full capacity, and keep everything you are working on damp (**A**).

Keep making wraps with the lapping, every time using the needle to create a channel through the straw to stitch through. Be mindful also to keep the needle at the same angle, catching the same amount of straw each time you insert it, aiming for repetition of technique. It can help to draw an imaginary line around the circumference of the rowl you are stitching through to aid this repetition. Perhaps two lines. One where the needle enters the straw, and one where it sticks out, helping you create the same angle and the same depth each time you make a wrap. Try to aim for some equidistance between them.

Now you have a new piece of lapping added, you can make several stitches before tightening them all up. Make one stitch, creating a big loop, but don't pull it tight. Then take the head of the lap and loop it round to make another stitch, so that you are working with one big loop, followed by one small loop (**B**). Take the needle out from underneath the small loop so it doesn't slip back out. Then tighten up the big loop, and the small one becomes the new big loop. Then make a new small one, and so forth. This means you won't have to finger and thumb along to the end of the lap anymore to make sure it doesn't twist, which is time consuming. It will speed you up, and you'll have to feed your feeder more frequently.

Keep going on your angle of elevation, repeating the angle of the needle, and not stitching too deeply, just catching fewer than a dozen straws. This continuity of placement and angle will create a symmetrical shape. You are aiming for the classic bell-shaped skep hive. It is the strongest and most iconic. Form follows function. The weight of the bees and the comb will be transferred down through the walls of the skep in a parabolic fashion. Remember to keep it wet, to create nice curves and prevent angular bends in the straw. Slightly twisting the straw in the direction that the lapping runs every time you tighten a stitch also helps it curve nicely.

A

B

You'll get into a rhythm. It can be a sit-down job working it up like a bowl in your lap, or you can stand at a work bench. Hold your work out at arm's length regularly to check it. A common mistake is to curve it too quickly too soon.

6. Working up the sides

When you reach a diameter of around 38–40 cm (15–16 in), you can start to put the needle in at a shallower angle until you are eventually placing it in horizontally. Then begin to raise the sides of the skep straight up, as it sits in your lap like a large bowl.

Try to repeat the horizontal stitches 'to a tee' for the rest of the structure, keeping the walls straight. Always catch the same amount of straw, with the needle going into the imaginary line you have drawn with your mind's eye around the rim and coming out at the other imaginary line. As you tighten up the stitches, the forces applied can often bring the wall too far inwards. Be mindful of this and hold the straw fast, exactly where you want it, as you tighten up to keep it in place, and raise the walls. For this middling-sized skep, you can raise it to be 40–43 cm (16–17 in) deep.

7. Finishing off

The easiest part of skep making is finishing off. In fact, one could say that it starts off difficult and gets easier and easier until you've completed your hive. When you wish to make your last rowl, take off the feeder gauge and carry on stitching until the straws taper down to nothing. You may need to add a few straws here and there to help it taper more smoothly if it looks like it is going to end abruptly, to help achieve a level base. This is your last rowl, so keep tightening the stitches and working any slack in the lap forwards. If it does look a bit uneven, you can soak the last rowl and whack it onto a sturdy flat surface or workbench to help flatten it. To tie up the loose end of your lapping, finish it off with a half-hitch or a cross-stitch (**A**) and tuck in the tail (**B**). it is then finished and ready to trim (**C**).

Alternatively, if you have any lapping left, you can keep stitching around the rim, which will take most of the wear and tear (**D**).

8. Making a door

A bee skep doesn't necessarily need a door in it, as one can be fashioned into the wooden floor if required. However, most of us love a fine entrance, and it would be remiss of me if I didn't teach you how to make one here. I often make a door in a skep, regardless of whether it is designed for simple swarm-collection, or for housing a colony. Charles Butler preferred his skeps to have a door in the lowest rowl – this should be worked into your final circuit as you come to the end of your skep.

1. For this Butler-style door, you'll need a 15 cm (6 in) section of willow rod. Split it evenly along its length into two pieces (**A**) and sharpen one end of each piece. Note that in this picture a finishing cross-stitch braid has not been added. It may be added before or afterwards or not at all.

2. Bearing in mind the door-post placement advised in Step 3, cut down through most of the straws in this final rowl, say 90% of them (**B**), and shunt the loose straws over. Continue stitching three times to bind the 10% of straws still intact (**C**). Then continue binding all the straws as usual for a few more stitches, being careful they don't slip out (**D**).

Making the door (cont.)

3. Take the sharpened sticks and hammer them into each side of the entrance, making two door posts that can take the weight of a full colony, preventing the lower rowl from compressing (**E**). Carefully align them so as not to break any lapping. To avoid this, I usually let them stand proud by a centimetre/half an inch and then saw them off level with the base (**F**).

4. After you have put the posts in, you can take the feeder gauge off and then taper the straw down to nothing as above, stitching around the rim with more lapping for protection. Keep everything wet as you tighten it all up as much as possible. Butler also fashioned reduced entrance blocks, which he called a 'winter doore' (see images on pages 50–51).

Letterbox entrance

To make what I call a 'letterbox entrance', keep the feeder on and carry on making more rowls to create the doorway higher up, cutting through around 90% of the straws, and stitching over those that remain. This important continuity of straw and lapping lends strength to the structure. You can then simply add one more rowl or keep going with many. You can crop the straws very close to the laps, making sure the lap nearest the entrance doesn't slip.

If you are not intending to keep bees in your skep, and simply want to collect swarms in it, Congratulations! You have finished. Look after it by keeping it dry. If you are planning to keep bees in your skep, there are a few more features you'll need.

Letterbox entrance.

Letterbox entrance in a skep sewn up with split green willow.

9. 'Sticking' the hive

To support the bees' waxen combs, which may press against each other during any movement of the skeps, sticks are inserted to help transfer their weight to the walls. There are many different traditional ways to do this. The simplest is called 'cross-sticking'. The sticks, often known as 'spleets', are fashioned from split rods. These are easily made from hazel or willow, dogwood or buckthorn, cherry or plum. The latter hard woods are of course more durable. Durability is a consideration when coming to remove them, as you may have to pull hard on one end with a pair of pliers.

Cut your rods slightly longer than the diameter of your skep and split them in half or into quarters depending on their thickness. A spleet needs to be slight, but strong. I usually split a rod the size of my index finger in half.

Sharpen one end of the spleets and insert them into the hive, in between the rowls rather than through the middle of a rowl (like swords in a magician's cabinet). A hammer may be useful here. I usually put in two spleets above the fourth rowl from the bottom, and again above the eighth rowl (**A**). Again, be careful not to skewer or snap any lapping in the process. I also usually put them in perpendicular to the entrance, as bees often draw their combs in line with the entrance. This is known as the 'cold way'. Sometimes though they will draw their combs the

A

B

Swarm in a spleeted skep.

C

Spleets supporting the combs.

'warm way', especially if they have a large entrance or are in a windy situation and need to buffer draughts and damp. Whichever way the bees decide to draw their comb, they are the architects of their own future, and will be designing their wax with their best interests in mind (**B, C**).

How comb forms around the spleets.

In *The Practical Bee-Master* (1780) John Keys wrote that in his opinion country people fixed too many spleets into their hives, which hindered the honey harvest, and resulted in the fouling and wasting of honey. He declared that one stick was enough, fastened about 5 cm (2 in) from the bottom, from right to left. At harvest time he would then cut through the combs as far as the spleet, remove it by giving it a wiggle and then pulling it up and out, and continue to remove honeycombs unhindered.

If you decide to employ this method, be very careful if you want to inspect the colony before they have fastened their combs to the one spleet. Whenever you are handling a skep, it is safer to roll the hive back slowly and steadily onto its top in line with the way the combs have been drawn so they don't fall together and squash bees. If you roll it back perpendicular to the combs, the weight of the upper ones may result in squashing others, and bees.

For as the Bees generally build their combs in parallel lines to the door-way, the spleet or stick being placed at right angles ... when the combs are to be taken out, a notch is to be cut in each comb as far as the spleet, which may be then pulled up, and the combs taken out entire.

JOHN KEYS,
THE PRACTICAL BEE-MASTER, 1780

10. Clooming

Once you've fastened the cross sticks or stick securely into your hive it is ready to be 'cloomed'. I am lucky to live on a farm with an organic herd of cows, producing quality cow dung largely free of the problematic chemicals and medicines that are damaging to invertebrates. This is a perfect coating for a hive and can be plastered on neat. Simply collect it in a bucket, stir it with a spade or a trowel and apply to the outside of your skep, rubbing it well in to the recesses in between the rowls (**A**). It will be made even better if you mix it with sand and temper it with wood-ash, and perhaps some chopped and soaked flax or wheat-straw. Recipes vary; it is commonly known by bee-masters and -mistresses as 'cloom' or 'cloome' and is similar to daub and cob. An application of a composite material like this on the outside of your bee skep will add insulation and damp-proofing, and deter birds and rodents from grazing the basket for nesting materials. Plaster it all around, and then set it in the sun to dry. A garden fork is handy to prop it up on to help it air all around (**B**).

You might think that cow dung is an odd thing to plaster onto a hive that you will at some point harvest a food product from. Rest assured, when it is cured it has antibacterial properties. Like a good wooden chopping board, if it hasn't been through a high temperature wash, it can absorb bacteria and thwart its activity. There is some wisdom in the old ways. I remember teaching a class saying exactly this, and a student revealed that they had grown up in India and that each week the floor of their house received a new coat of fresh cow dung. Which they appreciated, as it dramatically reduced the flea population.

11. Fashioning a base

Once your skep is dressed and dried, it is ready to be situated. Just like a cob cottage, as the saying goes it needs a 'good hat and pair of boots' to protect it from the elements: a decent roof and footings. Bee-skep floors are traditionally made of wood or stone. Wood is preferable as stone can become too cold in winter and too hot in summer. Ideally the floor is made just a couple of centimetres/an inch wider than the bee skep, with an alighting board or flight-board protruding at the front. The thicker the floor, the more insulation value it will have. I often chisel out a channel in the alighting board and slightly chamfer the edges. This rebate in the wood can itself serve as an alternative to an entrance created in the straw skep. This picture (**C**) has both a letterbox entrance and a wooden channel to act as a flight-board.

C

Forager bee returning home to a cloomed skep.

Set hiue [hive] on a plank,
(not too low by the ground)
where herbe with the flowers
may encompas it round:

And boordes to defend it
from north and northeast,
from showers and rubbish,
from vermin and beast.

THOMAS TUSSER,
FIVE HUNDRED POINTES OF GOOD HUSBANDRIE, 1573

Chapter 5

Making Hackles

The symbol of a busy skep, sitting in sunlight, surrounded by verdure and flora says many things to one's soul, yet it is misleading to the practical beekeeper. A bee skep would not be fit for bees if left a year unprotected.

The most charming and ancient shelter is known as a 'hackle' or a 'coppet'. This is a conical hat made of straw, and sometimes reed or willow. It acts like a small, thatched roof to shed rainwater and buffers the colony and skep from wind and cold, and the heat of the summer sun.

A skep also needs to be above the damp and potential predators, below the wind, and of course easily accessible for the beekeeper. Special stools and stands were made by old skeppists, some preferring three legs, others four, and some preferred one large sturdy leg fashioned like that of a bird table or driven into the ground like a post. To make a simple stand for a skep that is 40 cm (16 in) in diameter at the base, you'll need a piece of wooden board 23 cm (9 in) by 2.5 cm (1 in), and 91 cm (36 in) long, and some battens.

Charles Butler used 'hackles', which take their name from the ruff of long feathers around a cockerel's neck, or a dog's mane (when its hairs are standing on end in a defensive moment, its 'hackles are up'). This conical straw cap protects the skep very well from the elements; not just from winter cold and damp, but the summer rains and heat as well. The first time I made one of these hackles according to Butler's directions was a revelation. I've not seen any made in this way anywhere else other than a few old photographs. The result is satisfying and sturdy. I had made many hackles before, according to the stitched styles that Alston and other skeppists had perpetuated (shown later in this chapter), with the use of large needles and a lot of cordage. Butler's technique, however, has a simplicity and ancient craftiness about it.

A Butler-style hackle and gart (unfastened).

Making a Butler-style hackle

You'll need one good-sized sheaf or bundle of thatching straw.

1. Take out four or five handfuls, put to one side and soak well with water.

2. Clean the original bundle as much as possible, removing the leaves of the straws and any bits of dock or other such stems, as well as broken straws and those with dog-legs at their base that kick out sideways. It helps to tap the upright bundle onto the ground a few times and then put it down on its side. Sit on top of it compressing it with your knees and pull out the leaves from its base. Tap any straws sticking out back in, removing those that are weak or broken or for some reason won't tap back in. Now soak the bundle well with a hose and spread it out a bit upon the ground.

3. Feathering the straws up at the centre, arrange them so that all the heads are uppermost, forming a cone-like structure (**A**). It will take time to sift through the straws to find any smaller ones that may be hidden amongst the bundle and bring their heads up. Comb through with your hands and fingers to find any anomalies, rodent nests, bent straws and suchlike. The more time you spend preparing and arranging the straws, the better your hackle will be. Keep bringing straws up to the pyramid top, especially those with ears still attached.

4. With a good piece of cordage, tie your bundle tightly about two thirds of the way up and below any ears, and then if you wish you can stand it up to work with. Then give the whole thing a good shake and take out any straws that have come loose. Reintroduce them by slotting them in underneath the tight bind of cordage. After some time sorting, shaking and combing, reintroducing straws, taking off leaves and suchlike your bundle is ready for introducing some new straws to make the top cap.

A

5. Pick up a handful of the soaked straw you set aside earlier. Take off any ears and leaves. Then lay half of them one way and a half of them the other to create a small handful of even thickness and mingle them with the uppermost straws at the top of the cone.

6. Twist together the top of the cone and the new straws sticking out, to make a strong section of straw cordage. This is known as a 'litch'.

7. Bind the litch around the top of the cone in a tight circle (**B**). Keep introducing wet straws to the litch, as you would when producing a rowl, to keep an even diameter. You don't need a feeder gauge, just do it by eye. Don't let go or it will come undone. Keeping the tension, twist and then bind the litch as you continue to spiral, forming the head around the top of the cone, and working it downwards. Tightly butt up each round against the previous one, bringing any slack forwards as you work (**C**).

B

C

8. You'll need to continuously introduce new prepared straws as before to keep the rolled rope an even thickness. It's a good idea to have this straw prepared beforehand because you'll only have one free hand.

9. When you have reached the string, you can feather out the straw rope a bit to make it a slightly thinner diameter, then keep working tightly downwards past the string.

10. Lay the hackle on the floor horizontally and force the end of the rope in between the cap and the head of the pyramid using a forked wooden peg and mallet (**D**).

D

11. Once secure you can stand it back up and trim off any straws sticking out of the cap. Adjust any straws in the body, combing as you go, and inserting any strong straws that may have worked loose back under the cap.

Mounting your hackle

Your hackle is now ready to set onto your skep. It is like a solid broom of straw that you will need to splay out to fit neatly onto your skep. Your skep ought to be on a skep stand before fitting the hackle. Soak the hackle well if it has dried, and pick it up vertically. Bisect the straws by putting your arm horizontally through its middle and place it down centrally onto your skep. Splay and spread out the straws all around so there is even coverage on all sides, with the ends of the straws overlapping the edge of the skep-stand. Trim a section of the straws above the alighting board to give the bees a clear passageway. Imagine it's raining and try to position the hackle for maximum protection of the skep, shedding water away. The bigger and thicker the better, but if your hackle is small and the lower part of the skep is still visible underneath the skirt of the hackle, adjust it so that it sheds any rain beyond the edges of the skep stand. The damp is an enemy to bees.

Finally, you will need some briars or 'withies' of a pliant plant like bramble, willow, hazel, honeysuckle, bryony, or bindweed to make a belt hoop. This is sometimes called a 'gart', and old barrel hoops are useful for this too. If your skep, stand, and hackle are set in a particularly windy situation, it is advantageous to tie the hackle to the stand, by looping some cordage around the gart, under the stool, and back up again to hold everything fast until the weight of the colony gives it enough mass to hold itself. Similarly, one might weigh the hackle down by suspending stones from the gart with the help of a little cordage. Stones with holes in, known as 'hagstones' where I grew up, are perfect for this job.

A well-made and looked- after hackle will last from two to six years. If you like, the cap of the hackle can be cloomed from time to time, to give it a bit of extra weather proofing. I once cloomed some hackles all over, which gave them a very good weather-proofing, but made them incredibly heavy, and material would often fall off when I moved them.

Skep and hackle with a barrel-hoop gart.

A collection of hackles.

Making an Alston-style hackle

Straw covers fashioned in the style of more recent skeppist Frank Alston take a little longer to make, and may seem sturdier, but will still need repairing or replacing after a few years. Here's how you can go about it.

1. Clean up some good long straw wheat, soak it and lay it on a work bench. You can leave any ears of wheat on the top. Farmers will call those the heads, but thatchers will call them the tails.

2. Level up the butt ends of the straws on the bench and thread some good strong cord or bailer twine through a large needle.

3. Take a good handful of around 30 straws, roughly 5 cm (2 in) in diameter, and tie a tight knot around it, about 10 cm (4 in) up from the butt end. I often use a clove hitch for this. Leave a tail of about 23 cm (9 in).

4. With the rest of your cord and your needle, loop around another handful, making a round-turn that catches the clove hitch. Keep adding bundles of the same thickness each time, making a round-turn that catches the previous one, progressing along the work bench (**A**).

A

5. Measure the circumference of your skep with a tape or a piece of cord, and this will tell you when you have made enough hackle to wrap around it sufficiently (**B**).

6. Once you have reached your desired length, take up any slack in the cord and tie off with a knot, again leaving a tail of cord about 23 cm (9 in) long.

7. Repeat this whole process again about 60 cm (2 ft) up the straws, with another line of stitches, splitting the original handfuls in half, so that the new round-turns bind up a half of one handful to one half of the next (**C**). This means you need to start off by binding up a handful and half with the first hitch (clove hitch

B

C

or timber hitch). Note that the handfuls will be smaller because you are reaching the thinner end of the straws, and it will start to taper the whole form slightly towards the top as the second row of stitches tightens up, and the mat will become trapezoidal upon the work bench.

8. Once you are done, give it a good soak and position it around your skep, tying off the 23 cm (9 in) tails either to each other or into the stitches.

9. Then bind the hackle as fast as you can, close to the top, with your cord – a couple of times if you wish. An old terracotta plant pot is then the perfect addition to cap it off (**D**). Or you may even wish to employ the Charles Butler-style twisted litch-cap. Again, this style of hackle can be weighted down and tied to the skep stand using a gart if in a windy situation.

Hackle, gart, and mouse guard.

Other shelters

Probably the most well-known shelters are called 'bee-boles', and in a few places 'bee holes'. These are skep-sized recesses set into walls of stone, brick or cob, up to 76 cm (30 in) high and 71 cm (28 in) wide. The height of their base is most commonly around 46 cm to 76 cm (18 in to 30 in) above ground. Some have wooden shutters that can be closed in winter, and some are found in cellars to be used as winter quarters. I built some extra-tall ones next to a bee-house to accommodate a double-hiving technique that I'll explain later.

Other simple rustic shelters have been upturned creaming pans, pots, old sacks, and similar re-used objects. It was also once a common practice to site skeps in an open-sided shed or under a lean-to.

Right: **Swarm arriving at my home apiary bee-boles at Westmill Farm.**

Below: **Skep shelter with doors closed for the winter.**

Other Shelters

Above: **Skeps in bee-boles, with neatly fitting wooden floors.**

Above: **Bee skeps in an open wooden shelter, with very simple floors.**

Right: **One of the skep shelters at Westmill Farm, with a living roof.**

Chapter 6

Skep Beekeeping

Thus doth our Author. And not only thus;
But like his Bees, makes honey too for us.
And is contended that, to help us thrive,
We should partake the profit of his Hive.
For which (my share) I thank him: and for those
The Muses-Birds; whose nature here he shows.
And mauger such as will his Paines contemne,
The Muses thus, by me, doe honour them.

GEORGE WITHERS, 'TO THE AUTHOR', 1623

Above is a part of George Withers' poem 'to the author' in the well-received 1623 edition of *The Feminine Monarchie*. There have been a great many skeppists right up to this day – enthusiasts, writers, makers, and associations (I list some of these in my acknowledgements) – that have kept the traditional straw-skep alive. This is especially true in Holland, where the art of skep making and lipwork never lost a strong presence, having been championed through the recent turn of the century by the maker and author Toon Brecklemans.

The most recent beekeeper to have managed many bee skeps in Britain was probably a Mr B. Mason of Burgess Hill in Sussex. A short film entitled *Bee Sting Cure* made by British Pathé shows the fellow welcoming visitors to his skep apiary in 1949, where he administers bee-venom as a cure for rheumatism to those who have travelled from miles around. Approximately 50 large skeps are visible in open-sided wooden shelters stocked full of honey bees. The narrator tells us that

Setting a skep above bees in a hedge (see page 98).

Mr Mason believes he's reached a ripe old age due to enduring over 1,000 stings throughout his lifetime, and to eating honey, which he calls 'nature's elixir'.

Swarming

So, how does skep beekeeping work? Let's start with gathering a swarm.

> A hive of bees, as I've heard say,
> Said to their Queen one sultry day,
> 'Please your Majesty's high position,
> The hive is full and the weather is warm,
> We rather think, with a due submission,
> The time has come when we ought to swarm.'
> Buzz, buzz, buzz, buzz.
> Up spake their Queen and thus spake she –
> 'This is a matter that rests with me,
> Who dares opinions thus to form?
> I'll tell you when it is time to swarm!'
> Buzz, buzz, buzz, buzz.

W.S. Gilbert, *The Bab Ballads*, 1869

The ballad excerpt above is a good example of the common misconception that the matter of swarming rests with the queen bee, as a kind of ruler. Our current understanding is that the worker bees decide when it is time to swarm, most often due to overcrowding. They will stop feeding the laying queen, to reduce her weight and rate of egg-laying, and prepare her for flight. The moment of swarming is very exciting to behold, as the bees speedily issue, in a frenzy, out of the skep entrance. The sky fills with them darting in all directions yet staying as one cloud of insects. As they rise higher and higher, a direction of travel becomes apparent, and soon a branch of a tree will become the epicentre of the swarm. A cluster begins to build there, until all the bees are together again, often only a few metres from the mother colony.

Swarm-gathering

Whatever one's age, situation, or century, a swarm of bees that is shaken, brushed, ushered, or lured into a bee skep is generally happier there. They have a good organic shelter from any breeze, protection from the sun and rain, insulation from noise, and there are many places for them to get a foothold. They can easily spread

Tree-top swarm collection.

out, diligently creating more opportunities to fan their wings and fill the air with a special pheromone to call in returning scout-bees, and those lingering upon the bough, bothy, or bike-shed.

I find this moment of gathering the swarm to be one of the most euphoric within beekeeping and life itself. The weather is usually sunny and warm, and there is often the scent of blossom and birdsong in the air. The experience is fortified by the pioneering spirit of the bees, as their celebratory array of flight and excitement settles into a cluster – vibrant and full of purpose. To move them securely into a skep can be a delight and sometimes a challenge. They may be well knitted into a hedge, or high up on a far-out branch, or spread out in several clusters, or half buried under the bonnet of a tractor.

Once they are enclosed by that well-wrought and insulated bee skep, however, they love it. You may be settling a swarm accompanied by friends or family sharing the experience, all fascinated and bonded by a reverence for the honey bees that often calms one's movements and sharpens one's awareness. You may be in the garden at home or on the patio of a stranger, who has let you and your beekeeping experience into their home with a cup of tea and inquisitive awe, only too happy to have the unpredictable mass of honey bees removed from their apple tree after calling for help.

There are many ways to usher the bees into the skep you have provided them. It all depends on their situation, and there are many situations. In *The Feminine Monarchie*, Charles Butler justly declared that 'The manner of hiving is many … [it] can hardly be taught by precepts … [but is] to be learned by use and experience'. I'll talk about a few likely situations here.

Gathering from below

For the first, imagine a tree in a garden in full green leaf, sweetly blossoming in the merry month of May. A neatly clustered swarm of bees is hanging from the end of a bough about 2.5 m (8 ft) up. One would fetch a ladder and position it safely near the swarm, being careful not to knock the branch they are clustered upon, perhaps resting it upon a different bough if possible.

If using a step-ladder, this may be set up close to the swarm, but not directly underneath, and with a friend or a prayer to hold it steady. There needs to be room for you to manoeuvre underneath the bees with a skep in your hands. A large sheet of some kind will be very useful. I often lay down an old bed sheet, duvet cover, or dust sheet. You can place this on the ground directly beneath the swarm, and even beneath the ladder if need be. A pair of secateurs in your pocket will also be very handy to clear the way for your descent.

If you have numerous skeps, select one of a suitable size for the swarm. Don a veil or a bee-smock if that's your style and carry the skep up the ladder. Bees in a swarm are often agreeable, but not always. Hold the skep securely underneath the swarm. The dream scenario is that you can raise the skep, holding it upside down, to cup and enclose most, if not all of them. Give the branch one short sharp shake. The majority of the bees will fall into the skep, and some will fly back up to the branch.

Descend the ladder with most of the swarm in the skep, which will be much heavier now. Once on

The Nasonov scent

Swarms smell great. The scent of the fanning worker bees is an intoxicating mixture of floral nectars, verdant joy, and warm collaboration and praise. There could be a more poetic term for that olfactory fullness that fills the air, but beekeepers currently know it as the Nasonov scent, or Nasonov pheromone. The gland was first described in 1882 by the Russian zoologist Nikolai Viktorovich Nasonov, who thought it produced perspiration. A while later, in 1901, Frederick William Lambert Sladen of England first proposed that the gland produced a pheromone. For me it creates an air of euphoria and calm, that can aid the satisfying, stimulating and sometimes challenging moments of swarm collection.

Above: **A swarm settled into a skep. Some bees are still on the shrub; time to put the kettle on …**

Above right: **Bees fanning at the entrance.**

the ground, place it carefully on the sheet, upside down. You may have a mind to observe them for a while, to look for the queen and enjoy the sensuality of the moment. Hold the skep securely to stop it from wobbling, keeping it upside down, and be careful not to squash any bees. You will see them spreading out around the walls of the basket. Some may be fanning their wings with the delightful swarm-scent. This should be reassuring, and generally means they won't all fly up back to the tree. This also means that any scouts who are out looking for potential nesting sites find their way to the scent of the skep, which will be stronger than any residual scent on the branch. After your inspection and reverie, very slowly roll it over to stand it upright on the sheet, again being careful not to squash any bees. If the skep doesn't have a door, then prop it up on one edge using a stone or a branch, giving the bees good access. Another reassuring sign here is a line of bees on the rim of the skep fanning away avidly as a muster-roll for their retinue.

You can now usually put the kettle on, and by the time you have returned with a cuppa you'll have a good idea if you have a queen in the skep. If most of the bees have rejoined their sisters on the branch, then the queen is most likely to be there, and they have reassembled with her. If most of the bees are still in the skep, humming with satisfaction, and with a few still fanning at the entrance, you are sure to have a queen. There may be a small cluster of returning scouts on the

Above: **Spot the queen! As it can be a bit hard, she is circled here.**

Left: **Bees running up to join a swarm with a queen.**

original branch. You might like to raise the skep up a bit, perhaps setting it onto a stool or a garden table. Sometimes there may be two or more queens in a swarm, and the bees may divide themselves equally. If so, keep repeating the process until they stop returning to the tree and gather inside the skep. Or you could fetch another skep and settle two small swarms, with a queen each. A queen bee will be slinking around in the throng of bees, surfacing from time to time, constantly on the move. It takes practice and experience to spot her, but once you do, you'll be entranced by her elegance and charms. Then she'll disappear again into the throng.

If you are collecting a swarm at home, you can place the skep where you wish to site it, or shake the bees into another style of hive, be it a log-hive, a Layens or a Langstroth. Any scouts will either find it or return to their mother colony. If you are collecting the swarm from somewhere else, you may wish to wait until dusk to give the scouts a chance to rejoin the bees in the skep, before you wrap it up in your sheet and transport it home.

The whole situation described above may also be aided with a sharp pruning-saw or a sturdy pair of secateurs if the branch they are clustered upon is thin enough to cut and carry, and if you have the permission or the persuasion to remove it. You can then carry it, swarm and all, to the awaiting skep or hive and shake them in, or even place the whole branch in, for transportation.

Gathering from above

Inspecting bees clustered on a branch.

Another scenario might be that you are collecting a swarm that is enmeshed into a hedge, fence, bush, or tractor, and there is opportunity and space enough to position your skep above them (see image on page 90). Make sure it is secure and won't fall if it becomes unbalanced by the weight of a swarm clinging to one side of it. Then drum the fence or whatever they have settled into gently to encourage their ascent. Once a small group of bees start to walk up to the shelter you have provided and start fanning, the others will often follow. You can drum the fence or agitate the hedge some more from time to time if they stop moving.

Gathering using wild comb

Another possible discovery is that the swarm you are collecting has already drawn out some wild comb onto a tree and have laid it up with nectar and brood. Have an appropriately sized skep ready with a half a dozen spleets sharpened at one end. Using a knife or hive tool, carefully remove one of the side-combs, and place it into the correct side of the skep and skewer it with the readied spleets. Continue this process, shish kebab-style, making sure you replicate the order and position of the combs as precisely as you can to their original layout. Then place the skep the right way up as close to the initial site as you can, and any foragers will return to their new home.

Attracting a swarm

Another way of stocking your skep with bees is to set it up in its permanent shelter as a bait-hive, and wait for a swarm to find it. Your chances of attracting a swarm will be higher if you rub around the inside with some flowering broad bean-tops, hyssop, or thyme. Pinning a small piece of old comb or propolis on the inside will also get the attention of scout bees. Some beekeepers like to bait hives with lemongrass essential oil, and there are also artificial swarm-lure products available from beekeeping suppliers.

Probably the most surprising part of *The Feminine Monarchie* is some of Butler's advice for seasoning a hive, to make it attractive to a swarm. He has plagiarised this almost word-for-word from Edmunde Southerne's beekeeping treatise, saying that one should allow a hog to eat peas from an empty skep, and then liberally

Setting a skep above a swarm enmeshed in a fence.

Setting the skep near the fence, to attract returning scouts.

A swarm that has built comb.

smear around their frothy saliva, which the bees apparently love. One other traditional practice to lure bees, described by the Hungarian ethnographer Béla Gunda, is to use moss soaked in urine. If neither of these are to your liking, then rubbing with fresh bee balm, hyssop, fennel, mead, or cream will suffice.

It is a wonderful moment when a swarm of bees arrives of its own accord, and is considered a blessing or a good omen. If you are impatient, you can of course purchase bees, but they'll most likely be on moveable frames, and you'd be breaking a traditional beekeeping taboo. The finest bees for you are local bees. They'll be an 'ecotype' for your area and suited to your climate and seasonal variances. For example, Yorkshire-bred bees will usually have longer hair than those native to Somerset.

Looking for a queen in a swarm settled into a skep.

Settling an unsettled swarm

One other scenario might be that a hive is actively swarming, and all the bees are in the air, flying as one excited crowd. They will usually settle nearby, but not always. Sometimes a colony has already sent out scouts to find a new home before they all issue from the mother hive or tree. It becomes more pressing when a swarm that has been clustered on a branch for most of a long summer day begins to fly. They will be up and off to a new home, which could be some miles away. There are a few traditions that beekeepers have employed in this situation. Here follows a charm inscribed into the margins of the Lacnunga Manuscript, a late Anglo-Saxon compilation of charms:

> Against a swarm, take earth; throw it with your right hand under your right foot and say: 'I seize it underfoot; I found it. Lo, earth is mighty against every being, and against hatred and against enmity and against the great tongue of man.' And against that, cast dust over [them] when they swarm and say, 'Sit, victory-women, sink to earth. May you never fly wild to the woods. Be as mindful of my well-being as every man is of food and home.'

This has indeed worked for me in the past, and much of the practical magic I would say is in the casting of the dust up into the flying crowd. The swarm can soon settle, seeking safety, sensing the disturbance. Over time your words may differ, according to your deliberations as a bee-charmer. One contemporary way of settling bees in such a situation is by spraying them with water; I haven't tried this, but I assume it is with water from a spray bottle or even hose-pipe attachment. The most relaxing way of dealing with this moment though, and perhaps the most bee friendly, is to have already placed around and about the place a good number of bait hives that will lure a swarm, and trust that they have found one of them and are on their way to it. I suppose you could also simply wave goodbye and wish them well.

In his first-century *Natural History*, Book XI, Pliny the Elder described a practice known as 'tanging', advising that 'to cause a swarm of bees to settle, you must strike on bronze vessels'. This was widely practised throughout Europe up until the nineteenth century. It also served as an alarm call to alert your neighbours of the swarm and your possible trespass as you follow it to make a claim on the bees.

***The Villagers Charm to Lure the Bees*, unknown artist, *c.* 1800, shows the longevity of the practice of tanging.**

Skep husbandry

Once a swarm is settled or lured, one can successfully keep bees in bee skeps, whilst artfully ensuring healthy and happy colonies. The question that we ought to answer first is: why? Haven't we moved on? Isn't that an old-fangled thing of the past that rightfully petered out with our forebears? What about diseases and the honey harvest? Beekeepers are the biggest vector for bee stress and bee diseases, and a skep hive is many steps ahead of the rest in this respect. As mentioned earlier, many of these questions can be answered by perusing old manuscripts and treatises, which can give one confidence and courage, as can a good mentor. The true gnosis of skep beekeeping comes through the practices and from the bees themselves, the primary sources, and they may well end up teaching you much more than any books can in the long run.

> Wonderful houses built with wonderful Arte,
> royal treasures stored within in them?
> By skill and labour all are made.
>
> CHARLES BUTLER, *THE FEMININE MONARCHIE*, 1623

Mice

I had the pleasure of making a skep, cover, and stool for St Lawrence Church in Wootton St Lawrence, following the practical instructions within Butler's famous book. Like many other bee-masters who have written about beekeeping in skeps, Butler describes a variety of designs, capacities, and strategies, recommending some precise dimensions to follow for the fashioning of a hive. Of the two main styles mentioned, being 'wicker' or 'strawne', he prefers the latter, saying it provides better protection from mice, robbers, and wasps. I once had a group of mice chew their way into the back of a wicker skep hive. A small damp patch had appeared low down at the back after the rain cover began to leak a little and there the mice broke in, gnawing through the wet wicker. It had been rendered with a mixture of cow dung tempered with sand and wood-ash as Butler recommends: 'The best cloome for the hive is Neats dung ... but to harden it temper it with Lime or Ashes, with sand or gravell, which is good against the gnawing of Mice'.

Despite my repairing it with fresh mixture, the mice kept on digging and chewing their way in once they knew their reward. I repaired it every day, and in the morning, they had got in again. I eventually resorted to a mousetrap and caught half a dozen over the course of a week. Unexpectedly, that colony of bees did very well the next year! Possibly because they had drawn a lot of fresh combs after the mouse damage and put in a new coat of propolis. One strives to keep the mice

A propolis-preserved mouse

My first ever introduction to a beehive included the body of a dead mouse sticking out from underneath the hive roof. It was completely dried up and desiccated and its head was inside the hive, polished by the honey bees with propolis to prevent it from decaying. I imagine it may have been there since late winter, having found a chink in the armour of the hive. It had decided to explore the warm and humming home of around 20,000 inhabitants that time of the year, looking for food and some winter comfort. It's possible that the mouse had followed its nose and trails of dead bees to the hive. For, on a sunny day in winter, undertaker bees will carry out their dead sisters, lift them up in flight for a second and then drop them down, continually cleaning and cleansing their home within. The defensive bees had obviously sensed their intruder, and with their venom and vehemence, gave it what for. The bees have the power to kill an intruder with their poison but lack the strength to be successful undertakers for anything much bigger than a bumble bee. Any cadaver or potential health hazard that they cannot carry out of the hive, they will embalm with a coat of propolis, sometimes after removing hairs and offending items. Propolis is such a powerful preservative that it will prevent that body from decaying.

out all the same. I've never had any rodents chew their way through a straw lipwork skep, and it seems that the far more experienced Butler may not have either. He recommends looking under the skep-covers and hackles for pests like moths, mice, and earwigs.

Further advice from Charles Butler

> The best cover for Hives is
> a thicke hackle:
> *Alvearia stramento operiri*
> *utilissimum*

Butler quotes above from Pliny the Elder's magnum opus *Natural History*, 77–79 CE: 'it is very useful to cover the beehives with straw'. It is interesting to note that throughout *The Feminine Monarchie*, Butler never once uses the word 'skep', but consistently calls a basket for bees a 'hive', and in several places, he quotes Virgil's and Pliny's use of the terms *Alveari* and *Alvearia*. Today, a wicker-hive is termed an 'alveary'. He included much Latin in *The Feminine Monarchie*, some Greek, and a little bit of Hebrew, but gives no translations for the reader, who must translate it themself.

Butler's middling-sized skeps held three pecks (approximately 27 litres), and he taught us to have many of that size at hand, but also to have some larger and some smaller. It was common practice in skep beekeeping to match the size of a skep to the size of the swarm being collected. He also uses the term bushel, which is four pecks. He agreed with the philosophers that the sphere is the most perfect form, and the more spherical one makes a hive, the better. This suits the way bees hang their combs, and their cluster in wintertime. The reverend had a unique way of fastening spleets inside his hives. They worked together with a 'coppe', a circular piece of wood with a small central recess that is fashioned to fit in the apex of the ceiling of the skep. One end of a round rod is positioned into the middle recess of the coppe and then split three ways for a middling hive, and four ways for a larger, with the ends sharpened. These split pieces then arc outwards to attach the sharpened ends into the straw walls, a few rowls from in from the base.

Above: **Two freshly woven alvearies, waiting to be plastered with cloome.**

Above right: **Alveary with fresh cloome applied.**

Left: **Butler-style skep and hackle, St Lawrence Church, Wootton St Lawrence, UK.**

Below: **A Butler-style coppe and spleets arrangement.**

The advantages of skeps for bees

Through the practice of skep beekeeping, it becomes clear that there are many advantages to the honey bees over those in more conventional hives. They are very well insulated in a curved cavity where there are no 'cold corners' that can occur in poorly insulated, boxed hives. This limits damp and condensation. As does the ability to varnish more of the nooks and crevices with propolis, which has an anti-fungal and anti-mould action. More importantly, propolis is one of the strongest natural antibiotics known and plays a vital role in protecting the colony against pathogens. Bees on moveable frame hives use less propolis for two main reasons – they are most often made of sawn and sanded timber, with far fewer cavities to fill, and the wooden frames give most of the strength to the combs. In a fixed-comb environment, bees mix a lot more of the vital propolis into the wax to strengthen it. The Swiss bee-scientist Francois Huber termed these as 'strong pillars' and noted a lack of these in frames with a framed foundation. Within an undisturbed skep the bees are more able to maintain the propolis envelope, hermetically sealing themselves as well as possible. The agglutinant propolis works many wonders for a hive, including having a radio-protective activity, warding the brood and honey from the harmful rays of the sun. Interestingly, propolis from a mountainous area has a higher radio-protective action than propolis from a swamp, and the latter has a higher anti-fungal and anti-mould activity.

Propolis curtain being drawn at a skep entrance.

Bees are less stressed in the safety of skeps. Their combs are not removed or manipulated as much as they are in moveable frame-hive situations. This means that a lot less wax and propolis is broken, which they would spend precious time and resources repairing in the more conventional hive. In a well-protected skep hive, the bees can maintain the desired temperature and humidity that they work hard to create. If a colony is opened and exposed too frequently, this carefully managed environment is compromised. Given the space and opportunity, honey bees design and manage the internal dynamics of their colony in their very best interests. They can orientate the combs the way they want them according to their environment, make the cell-size they desire, build an undisturbed propolis curtain at the entrance, and keep the colony a healthy place in which to raise brood and successfully store honey and propolis. In a skep, the bees are essentially allowed to get on with their well-ordered plans and adapt them according to changes in their environment. A dynamic equilibrium exists between the bees and their environment, and the beekeeper can be a sensitive part of that environment. When given the privacy and ability to maintain the hygienic home they desire, bees are more resistant to disease.

Communication is an essential part of a colony. Taste and smell are a large part of honey-bee perception, and the hive scent and chemical communications are disturbed far less within a skep hive. Their hive scent is not dispersed as much as it is when one opens a moveable frame hive. Skep bees can control ventilation and air-flow more consistently in accordance with the way they have built it. Their dark olfactory temple is more able to function as nature intended, and not exposed to light and pathogens as frequently. Hence, the creation of the term 'Nature-Based Beekeeping' by Dr Nicola Bradbear MBE, and its adoption by Professor Tom Seeley.

Bees in skeps are less likely to be manipulated, alarmed, and subject to the pressures and strains of more intensive and manipulative styles of beekeeping that can compromise their immune systems – and therefore thrive as treatment-free colonies. A skep beekeeper makes minimal inspections, squashes fewer bees, and aggravates a colony far less. I have said it before and I'll say it again, that honey bees don't really mind what they are living in as long as it suits their needs, and it is the spirit with which they are kept that is most important. Hence, it is important to say that nature-based beekeeping is certainly not exclusive to skeps and other fixed-comb hives. It can be practiced in most styles of hive that one might manage treatment-free, and much better so if they are free to swarm. Bees dwelling in a skep though, or in another well-made fixed-comb environment, cannot have their plans as easily messed up by the beekeeper. They are the architects of their own future, something they've had over a hundred million years of experience practising.

Working with swarms

Skep beekeeping is not all about settling a swarm into a basket and letting them get on with it, though. There are some tricks to the trade and a few ways to help them in their work, and yourself in your own. Anticipating swarms is one of them. The bees that I look after are all free to swarm. I am lucky to live on a farm with no immediate neighbours whose chimneys may be coveted by bees looking for a home. Much of contemporary conventional beekeeping employs methods of swarm control. These might involve weekly inspections, looking for queen cells with eggs or larvae developing, moving frames around, splitting colonies, re-queening with a queen from another colony, clipping a queen's wing and more. In a skep, if one is keen to prevent swarming then there is a tried and tested technique. This involves driving the bees and is dealt with in the next chapter.

Swarming is a life-affirming moment. It is a part of the way that honey bees reproduce and increase, creating two viable colonies from one strong one. It is also one of their strategies for outsmarting pests and diseases and is a sure sign of health and vigour. An early swarm ensures a brood-break in the summertime that can be crucial for their own pest control. A brood-break is a period of up to four weeks when no eggs are being laid, and when any sealed brood will hatch out, leaving the cells open for cleaning by the diligent workers. This prevents a buildup of the parasitic *Varroa destructor* mite population, diseases that they are vectors for, and other disease agents that can proliferate inside brood-cells. The Varroa mites undergo their mating rituals inside the sealed brood, living on the larvae. Once there is a break in larvae developing, the mites must exist on adult worker bees throughout the colony. In treatment-free or wild colonies, workers will groom each other, and kill and remove the parasites.

This downsizing, emptying, and cleaning process is age-old wisdom. Think how our human nomadic ancestors used to move from place to place, or between summer and winter camps, leaving behind middens, pests, predators, and potential disease build-up; and also reducing the consumption of localised resources. Only half of the bees swarm off, and many will stay behind to continue to rear the brood, forage for food, guard the entrance, ripen nectar into honey, and see to all the business as usual once the swarm has flown. They will not have a laying queen present, but several virgin queens, ready to hatch. They do have stings but will only ever sting another queen.

The first virgin queens to hatch often fight, and the strongest or most skilful will win. They may also seek out the unhatched queen cells, which look a bit like waxen peanuts suspended from the combs, and dispatch their royal sisters by stinging them through the cell wall. The worker bees will then clean up afterwards.

Inspection

Observing the skep entrance

A simple and enlightening first step would be to observe the skep entrance. Whether you have a channel cut into the floorboard for the bees to alight upon and crawl inwards underneath the skep, or a letterbox entrance somewhere further up, a great deal can be learnt watching and listening to the bees come and go.

For example, if pollen is being brought home then it's a sign that they have brood or are expecting to have brood. If lots of pollen is coming in then one can discern that there is a lot of brood, or a well-mated queen beginning to lay. If there are yellow or brown spots of faeces then there is dysentery, and a possible nosema. If one hears piping or tooting, then secondary swarms are preparing to fly. If a colony goes still and silent in good weather during a nectar flow, a prime swarm may be about to issue forth. The behaviour of bees and any debris at the entrance may also tell you of queenlessness, robbing, Chronic Bee Paralysis Virus (CBPV), mouse activity, starvation, and more. A gem of a work solely dedicated to this subject is *At the Hive Entrance: Observation Handbook* by H. Storch. Another is *Look, Listen, Learn at the Hive-side* by Ron Brown.

Teaching skep beekeeping.

Hefting the skep

A second step you may like to employ is hefting the skep, keeping it on the stand and board so as not to break the propolis seal. To heft a skep, or any other kind of hive, take a secure hold of one side of its base or floor and lift it up, tilting to one side, giving you an idea of half its weight.This art will take some time to master, and for the wisdom of one's flesh and sinew to calibrate and assess the weight of the colony. A particularly light colony late in the year may be dwindling, queenless, and certainly have a dearth of stores. Hefting a weighty colony is reassuring, especially during the winter months when the colony is relying on what it has stored inside. (The kinaesthetic and sensitive tactile qualities that will be fostered within you may be beneficial in many areas of life in this increasingly abstract and mechanised world.) One could of course heft a hive on one side using a luggage scale and then double the reading to get a rough idea of the weight of the colony, once the weight of the skep and stand have been subtracted.

Inspecting the skep floor

A third step of inspection could be removing and examining a tray beneath a mesh floor (if you wish to use one). Upon this tray there'll be a wealth of information about what is happening inside. You will see, for example, where the bees are uncapping stores and brood, where they are building new comb, various pollen loads, Varroa, or evidence of hygienic behaviour (chewed Varroa mites, white antenna, or nymph parts). The late Ron Hoskins and the Swindon Honeybee Conservation Group have pioneered much of this work. A tray beneath a mesh floor can also lure the wax-moth to lay eggs there instead of the colony above. I used to put open mesh floors into skep stands in my early beekeeping days, so that Varroa mites would drop through, but no longer do as I quickly learnt that the bees can deal with the mites adequately. If you don't like the idea of a mesh floor, but are still keen to inspect any debris fallen from the combs, the standard skep stands or floorboards can be temporarily changed for clean ones, and the debris on the old one can be examined. I recommend putting the old floor back under afterwards as it will be polished with propolis, and will fit well. Make sure you reseal the skep to the floor with a bit of fresh organic cowdung if required.

Upturning the skep

If what you observe so far is cause for alarm, you may wish to upturn the skep and look for queen cells or evidence of swarming; for a healthy brood check, one can part the combs and if a good amount of healthy sealed brood is present, fears may be eased. As Frank Alston wrote, 'a good slab of solidly sealed brood is a fair indication that all is well'. You'll be able to count the number of plates of brood-comb, assess honey stores and perhaps spot a queen.

The nature and frequency of exposed inspection of skeps is a double-edged sword. On one hand, the less you inspect the better. As I have explained, the bees are less disturbed and the pheromones at the heart of their world remain happy ones, propolis is not broken, their temperature is not lowered, comb is not broken, bees are not squashed or smoked, bees are not stressed or put at risk of disease entering their dark temple of sound and smell, taste and touch. On the other hand, you'll be less aware if a disease has entered their world.

An upturned skep ready for inspection.

Foulbrood

Ultimately, the worst-case scenario is that either you or a nearby beekeeper identify a foulbrood in a colony of bees. It is less likely that a well-kept and unstressed skep will be infected, but if there is a local outbreak, you have a legal and moral responsibility to make sure, and to report disease if you find it.

The skep would then need to be checked for foulbrood by your local bee inspector. To make a full inspection of a skep easier, it is well worth driving all the bees out into an empty skep first. See 'driving the bees' in chapter seven. The combs can be cut out and inspected one by one as the cross sticks or 'spleets' are slowly removed. A skep comb-cutting knife that is usually used for harvesting is a good tool for the job. It consists of a blade that protrudes at a right angle from the end of a long handle. The combs can be carefully put back again using new spleets if no sign of disease is present. The bees wouldn't thank you for it, and the bee inspector probably wouldn't either, but the job can be done.

With due care and attention, healthy combs can also be transferred from one place to another when checking for foulbrood. A few years back, a friend and I removed a colony from a roof that had been in there for several weeks. We simply took the combs out one by one from the closest end, inspected them and fastened them into an empty skep with cross-sticks. The straw skep had a flat top with a 7.5 cm (3 in) feeding/expansion hole in the top. The colony was fed and built up well. This colony didn't actually have any signs of disease, so we were able to transfer all the combs.

There are two types of foulbrood that effect the bees' larvae and pupae and can weaken and kill colonies: European Foulbrood (EFB), caused by the bacteria *Mellissococcus pluto*; and American Foulbrood (AFB) caused by the bacteria *Paenibacillus larvae*.

European Foulbrood is linked to stressed colonies – especially in a spring build up or in hives that are frequently moved, a larva will contort in the cell and lose its pearly white healthy quality, becoming dull, cream, brown and eventually green. It has a sour smell. Healthy, unstressed colonies are thought to either not succumb to the endemic EFB or to quickly overcome it. If a colony is found with EFB it can be saved by various treatments and by removing the infected combs.

American Foulbrood, however, is more virulent and deadly, being spread by spores germinating in the ventriculus of a larva, which may eventually cause it to die of septicaemia and completely melt down in the capped cell. These become sunken, moist and discoloured. A 'matchstick test' is used in identification; as one probes a sunken cell, a sticky ropey substance will be drawn out. In advanced infections, some report a smell akin to foul dead meat. A colony with AFB must be destroyed, hive and all. Honeycombs will have spores too and can infect any bees robbing a weakened or collapsed colony.

One should never feed honey from an unknown source to bees due to the risk of the spread of disease. Old hive parts and equipment and old combs can harbour disease spores too.

Skep beekeeping can be a wonderful addition to more conventional styles of beekeeping, as well as in stand alone situations. Either way it is wise and educational to join your local bee groups and associations, brush up on your bee diseases and become a part of the vibrant social dimensions of bee crafts.

Apiary hygiene

Limiting the spread of disease is one way a beekeeper can help bees. Robbing can occur between colonies in times of dearth, especially in the late season when the nectar flow slows down and the days grow shorter. Wasps buzzing around hive entrances will also be exacerbating this from early August onwards and can deplete and kill off a weak colony. Diseases and pathogens can be spread when bees rob each other. Naturally, honey-bee colonies tend to be more spread out, but within a conventional apiary, colonies are situated closer together for convenience. If a beekeeper spills honey, then any pathogens that might be present can also be picked up by scout bees.

Far left: **Wooden mouse guard.**

Left: **Skep made in a French style with a square wooden rim and mouse guard.**

Below left: **Metal mouse guard on an alveary skep.**

Endeavour to keep the apiary clean and tidy of spilt honey, old combs, and exposed hive parts that may encourage colonies to rob each other. If a sudden source of honey arises nearby, say from a spillage, scouts will find it and report back to their hive – not by making the famous figure-of-eight waggle dance used to communicate precise distance and direction of far-away forage, but with a round dance to indicate the nearby source. Without any direction, a frenzy of bees will soon ensue as hundreds will be out searching every other hive entrance, nook, and cranny. This behaviour, and the presence of wasps, will greatly change the temperament of honey bees.

Hygiene is especially important for skeps – and any fixed-comb beekeeping – because the brood-combs cannot be as easily scrutinised for signs of disease as they can in a moveable frame-hive. Signs of the notifiable diseases European Foulbrood and American Foulbrood are not so easy to spot in skep beekeeping. The spores of the latter, more destructive variety, can live up to 40 years. Renewing combs often is one way to reduce any build-up of disease on old, dark brood-combs. It is tempting for a beekeeper to keep a hive or skep full of old combs after a colony has failed, whether that is due to queen failure, or continuous swarming. However, in the long run it is far healthier to cut out all the combs, and you even can scorch round the inside of a straw skep quickly with a blow-torch to kill off spores and pathogens, as one might when cleaning a conventional wooden hive. Have a damp cloth handy just in case any straw catches. These combs can then be rendered down to make clean wax and any beeswax products you may wish to produce.

It can be helpful and reassuring, certainly for a bee inspector, to keep one or two moveable frame-hives that you can easily inspect in or near to the same apiary as your skeps, as a kind of control, especially if you are in an area where foulbrood outbreaks are common or recent. Bees in skeps are free of many of the stresses and strains that bees kept conventionally are put under, and a die-hard skeppist might see this as a chink in the armour of the skep apiary, as beekeepers are one of the biggest vectors for bee diseases. So, these could even be lipwork top-bar hives (I feel another book coming on…), as mentioned by Thomas Wildman in his *Treatise on the Management of Bees* (1768), or what has been called a 'Greek Hive' or 'Pompeian Hive'.

Keeping equipment and clothing clean will also help limit the spread of disease. Spreading out colonies, too, will benefit them greatly if one has the space to do so. Mice, ants, and hornets are another possible source of pathogen transference. Even if a skep is well guarded from mice, with a simple mouse guard pinned to the entrance, they may still build a nest on top of it in the winter months. It's good practice to check under hackles or skep-shelters from time to time through the darker half of the year.

Chapter 7

The Harvest Home

Go gather with skill,
and gather that will.

Driue hiue, good conie,
for waxe and for honie.

No driuing of hiue,
till yéeres past? Fiue.

Good dwelling giue bée,
or hence goes? Shée.

THOMAS TUSSER, *FIVE HUNDRED POINTES OF GOOD HUSBANDRIE*, 1573

The quote above is from a list of late-season tasks in Thomas Tusser's poetic book on good husbandry. Today, all sorts of learned bee folks often declare to me, regarding skeps, 'well you can't do that any more, can you? Because you have to kill the bees to get the honey out'. This is simply untrue. I have seen similar misinformation written on beekeeping websites, in contemporary magazine articles, and even in brand new bee books.

If there is one resounding and verifiable piece of information that I'd like you to take away from this book, it is the knowledge that one can keep bees in skeps, ensuring their health and vigour and making a honey harvest without harm to the bees or to the skep. When one looks further back to the likes of Pliny, Virgil, Aristotle, Varro, Palladius, Didymus and Columella there is no talk of killing bees

A 'super skep' or 'cap' full of honey harvested from a double hiving situation.

to ease a harvest. On the contrary, there is much evidence of skep beekeepers and fixed-comb beekeepers preserving and replenishing their colonies. They prized honey that was harvested without the use of smoke, left adequate stores for the bees in autumn and strengthened weak colonies by adding ripe brood or by uniting them.

There came a time when culling bees to make the honey harvest easier was popular in England, but it was never the only way. It was the preferred method of old bee-masters like the Revd Charles Butler and Edmunde Southerne. Butler describes several ways to harvest honey. Only one of these methods involves culling bees and brood, by suffocating them with smoke over a pit, to make the cutting of the comb easier. This seemed to be his go-to method but was not the only one used at that time. He was essentially a small-scale bee-farmer, and I've often tried to empathise with his apicidal tendencies.

During that era mother nature's bounty was seen as an inexhaustible resource given to humans by a divine being for their use and prosperity. I imagine beekeepers back then may have viewed a large apiary the way a forester views an area of woodland or copse: knowing that one could cut, coppice, and harvest here and there and it would all come springing back the next season. Butler was also writing at a time before the Enlightenment, bills of rights, or any well-established human rights in Britain, let alone any established animal rights in England or the Western world.

It was common practice for all who were in-sourcing a good part of their diet to cull some livestock at the onset of winter, and whatever creatures they may have fattened through the year to keep the life in themselves, their families, and communities. Butler had a lot of hives, and I also imagine that his culling ensured empty hives for the following year's swarms. It was also a way of reducing disease, as one would cull the heaviest (for the weight of the honey) and the lightest, which would be dwindling, and may not make it through the winter, and may be harbouring pathogens. Colonies of honey bees are still culled today by law if they have American Foulbrood. To reiterate, in this day and age, I see no reason at all to kill bees in order to harvest honey from a skep.

When to harvest

Virgil suggested twice a year, first with a spring harvest when the Pleiades are leaving us, and later as an autumn harvest when the asterism returns, reborn and rising in the constellation of Taurus. Arcturus, the 'Bear Guard', was also another signal star for harvest. Sirius, the 'honey-star', was said to help produce the finest honey.

If you're not a keen astronomer, St Bartholomew's Day, 24 August, was another traditional time to collect honey, or just before that, so you'd have some to sell at St Bartholomew's Fair.

Checking a cap of honey to see if it's ready to harvest.

Contemporary beekeepers start making a harvest at the beginning of August and usually finish by the end of September. Some take an earlier harvest at the beginning of June if there is a lot of oil-seed rape growing in their area, as its nectar sets incredibly quickly.

Whenever you are taking honey from a skep, make sure it is capped honey, sealed with wax, and not open-celled nectar, as this is unripe. Its moisture content will be too high, and it will most likely ferment. Make sure also that you leave the bees with enough honey to see them through the winter, or indeed through a wet spring. Hefting the hive to assess its weight of stores is a knack you will learn over time. It is best to err on the bees' side, of course.

If in doubt, leave it until the springtime, around the time of the Ides of March to the Vernal Equinox. In *Rerum rusticarum libri III* (37 BCE), the author Varro warns beekeepers of being too grasping, and encourages them to leave fallow years, and remove old combs.

Another time to harvest is of course when a colony becomes queenless, swarms repeatedly down to nothing, or becomes devoid of bees. The whole skep can be cleaned out, harvesting honey and wax, and as discussed above under 'Apiary hygiene', you can even give it the once-over with a blow-torch to thwart any disease, being sure to be quick and to have a damp cloth handy.

How to harvest

Harvesting honey from a skep is incredibly simple. When I'm asked how I do it, I'll sometimes nonchalantly say, 'I cut it out with a knife'. Which is true. The simplest way of harvesting honey from a bee skep is by cutting a piece of the honeycomb out from the smaller side combs. There is usually brood in the larger central combs; adjacent to that is pollen; and the outer combs contain honey. A cottage beekeeper with only a few colonies would simply take a piece of this when the honey-pot in the larder or the medicine cabinet was empty, using methods of extraction that didn't involve culling bees with the use of smoke from Brimstone or Puffballs, but those which kept them alive. This was termed 'exsection' or 'castration'. I suppose what someone might really be asking is, 'how does one get the bees out of the way before cutting out the honeycomb?'

There are several ways to do this. If one is planning to take only a couple of combs, they may be carefully cut away after any bees have been ushered off with a soft bee-brush. Some old masters used to use a feather, a goose-wing, or a frond of a soft plant like fennel. Charles Butler was not a fan of using wings or feathers as he found that bees often stung the central quill part, causing some to die and others to rally to the alarm pheromones released. You'll be the best judge of the temperament of your bees.

Removing spleets from a brood-skep.

The best time to harvest is of course on a warm and sunny afternoon. You may want to use smoke, and if I do I like to use dried tansy in the smoker. Spleets may need to be removed, and a pair of pliers will be useful for this job. Alternatively, one may cut around them. Be warned, though, that this may result in a stickier manoeuvre if you are cutting through honeycombs, which may in turn risk the bees becoming stuck in honey, and possible robbing by nearby colonies or wasps. If any bees do become besmeared with honey, they can be placed in a sieve sprayed with warm water and put in the sun to dry off. The cleaner and more seamless you can make the whole operation the better. Try not to crush any combs or drip any honey. It's a good idea to plan ahead and have everything you need ready, including your honey processing set up when you return to the bee shed or kitchen.

Driving the bees

Another technique used to help the harvesting of honey from a skep is known as 'driving the bees'. The Thomas Tusser poem at the beginning of this chapter is thought to be the first written account in the English language of driving bees from a hive and preserving them, rather than culling them. This involves upturning a colony and pinning an empty skep to its open base with wooden skewers or driving irons: two iron staples and an iron pin.

The upturned lower skep containing the bees and comb is then rhythmically drummed upon. This is best done a few meters away from the original stand. The

heavy skep of bees may be supported by a bucket, chair, or large plant-pot. For the first few minutes of slow rhythmic drumming, some bees will fly about, others will descend to fill their honey stomachs, and soon a handful will start their orderly walk upwards to the 'lifeboat' skep above. The bees' instinct in this upside-down state of emergency is to walk upwards to the shelter you have provided, away from any harm, wet weather, damp, light, or potential predators to re-group in a swarm-like cluster.

Driving the bees.

Once the first handful of pioneer bees has ascended and they have started to fan their wings, dispersing their 'muster here' Nasonov pheromone, it will usually take about 20–30 minutes for the rest of the colony to join them. The top skep containing the bees can then be placed on the original stand for any foragers to return to, and the honeycomb can be removed from the driven skep.

The bees can then be united with the mother skep and the brood afterwards by placing the skep full of bees on top of the still-upturned old one and giving it a quick, sharp pat. Alternatively, the bees can be shaken onto a board or sheet fastened to the front of the skep stand once the original skep is placed back up the right way on the original stand. They will then walk back up into their home.

Double-hiving

Another way of harvesting honey from a skep is by employing a double-hiving situation. After a swarm has been settled into a 'brood-skep' with a hole in it, and they have filled it with comb, a smaller cap (also known as a 'super-skep') may be added on top. All these skeps are usually flat-topped, having an 8–10 cm (3–4 in) hole in the middle of their crown. This hole is kept closed with a wooden lid or a small round lipwork mat until the super is added. The queen will tend to stay in the lower skep and continue to lay eggs there. The colonies will stay relatively small, as they will be swarming, or are a settled swarm from the present season. Hence, they will tend not to build any brood comb in the super. The adding of super-skeps is akin to conventional beekeeping in the sense that one keeps the brood in a lower skep and aims to harvest honey from the upper skep or skeps.

I was demonstrating this to a group whilst teaching skep beekeeping here in the apiary for the charity Bees for Development. We inspected a double-hiving colony and removed a cap which was full of honeycomb in early August (see top image on page 125). We then drove the handful of bees present in the cap into an empty, spare skep and returned them to the original brood-skep before removing the honeycomb. The brood-skep is traditionally called a 'depriving hive'. The super-skep is often called a 'cap' or a 'storifying hive'.

Brood-skep with a 10 cm (4 in) hole in the top.

In *The Bee-Keepers Manual, Or Practical Hints on the Management and Complete Preservation of the Honey-Bee* (1870), the author Henry Taylor details his work with straw skeps. He encourages beekeepers to use straw 'depriving hives' that are 35 cm (14 in) wide and 18–20 cm (7–8 in) deep (inside measurements) with an 8–10 cm (3–4 in) hole in the top. He credits these dimensions to J.H. Payne's *The Beekeeper's Guide* (1838). Taylor covers the central hole with a round mat of lipwork, pinned or weighted down. After the colony has filled the depriving brood hive with wax combs, the mat is removed from the hole and a straw super of the same dimensions is placed on top. In a good year with merry weather and an ample nectar flow, one can add two more straw supers, like a wedding cake. Taylor also adds a circular wooden board in between the skeps for added strength, of the same diameter and with the same sized hole in the middle. The supers can be less in diameter, around 30 cm (12 in).

Left: **Circular wooden boards for multiple hiving, Henry Taylor-style.**

Below left: **Double-hiving in bee-boles, no cow dung (the bees have sealed the join on the inside with propolis).**

Below: **Wooden boards inserted between each hive (in situ they will be sealed with organic cow dung).**

In *The Handy Book of Bees*, first published in 1870, A. Pettigrew champions the technique of using similar flat-topped skeps, with a 10 cm (4 in) hole in the top. He encourages readers to use large skeps, up to 50 cm (20 in) wide, building up to that capacity each season from smaller skeps by putting prime swarms into a slightly larger one each time until this maximum capacity is reached. He was well connected in the bee world of his time. To assess his theory of practicing a better beekeeping by increasing hive sizes he made a countrywide survey of the dimensions of his friends' skeps, when they first swarmed and their average

Above: **Simple double-hiving situation, sealed with cow dung, without boards.**

Right: **Smaller cap or super-skep on a brood-skep.**

weight at the end of the season. His largest skeps were similar in volume to a standard national brood-box today, and much better insulated of course. He stated that 'Straw hives, well sewed with split canes or brambles, are incomparably better for bees than any other kind of hive', adding that 'Hives made of wood, at certain seasons of the year condense the moisture arising from the bees'. On swarming, he says, 'The swarming system of management is not only more profitable, but, taking a run of years, is better every way, and more natural, than the system that prevents swarming'.

Making a flat-topped skep

1. To start off a flat-topped skep using a wooden ring, first find a suitable piece of wood around 10 mm (0.4 in) thick. You'll need a compass and pencil, a drill and drill bits, a vice or clamps and sharp chisels and gouges or a jigsaw. Mark two concentric circles with a compass, one at 100 mm (4 in) diameter and another at 140 mm (5.5 in). Cut out the middle hole starting off by drilling holes and then chisel or jigsaw the rest out. Cut to the second circle with saws and chisels, leaving you with a flat wooden ring. Then drill approximately fifteen equidistant holes around the ring with a drill bit of a similar diameter to the lapping you are using, most likely around 5–6 mm (0.2–0.24 in).

2. Next, take some well cleaned and soaked straws, and wet lapping. Begin using a half a dozen straws, binding them to the side of the ring by stitching through the small holes. Each time you make a new stitch, add another half a dozen straws (**A**).

3. When you have made around eight stitches, put your feeder gauge on to help hold them. This will eventually keep your rowls at a 40 mm (1.6 in) thickness as you keep feeding every time you make another stitch until it is at full capacity.

4. When you have stitched through every hole around your wooden ring, begin to use your needle to make the second rowl as with conventional skep making.

5. As this is a flat-topped structure, keep the needle going straight through the straw at ninety degrees, passing under one of the laps on the previous rowl. You are making a flat circle like a round mat. Remember to keep feeding straw when needed and to employ the rule of thumb.

6. When you have reached your desired diameter (35–40 cm/14–16 in), you'll need to abruptly change the angle of the needle to turn the corner and start to bring the walls straight up. To help at this point, make sure the straw is well soaked using a spray bottle or wet cloth. Now lift up your rowl of straw and place it on top the last rowl, and start to put the needle in horizontally. You will now be building the walls upwards, continuing to place the needle in horizontally. Take care to keep the walls coming up straight as you build your skep upside down (**B**).

A

B

Uncapped honeycomb

Be sure not to harvest honeycomb until it is properly capped. Uncapped honey will not be ripe, having too much water still present in the nectar for long term storage. Leave it another week or so for the bees to ripen, thicken and cap off with wax; or leave it for the bees to enjoy over the winter and make a springtime harvest. Any uncapped honeycomb that is harvested will absorb even more moisture from the atmosphere and begin to ferment. If you do take out a comb that is mostly capped, and you didn't notice the uncapped parts, you can cut that off and feed it back to the bees or put it into your mead-vat (see Chapter 9).

If you are going to feed uncapped comb back to your bees, do it at dusk when they have stopped flying so you don't encourage robbing. Needless to say, make sure you give it back to the colony you harvested it from. In a double-hiving situation, the bits of comb can be placed directly on top of the hole in the brood-skep, with a cap put back on top of it, and sealed all round with dung. The bees will clean it all up and you can remove it a few days later. In a single-hiving situation, you can place the comb in a shallow dish, lift the skep and put the dish onto the skep floor, under where you made the harvest, making sure that none of it will seep out of the sides of the skep. Any honey that leaks out will cause robbing by neighbouring colonies and attract wasps and other pests. When feeding skeps, it is a good idea to reduce the size of the entrance, or attach a mouse guard.

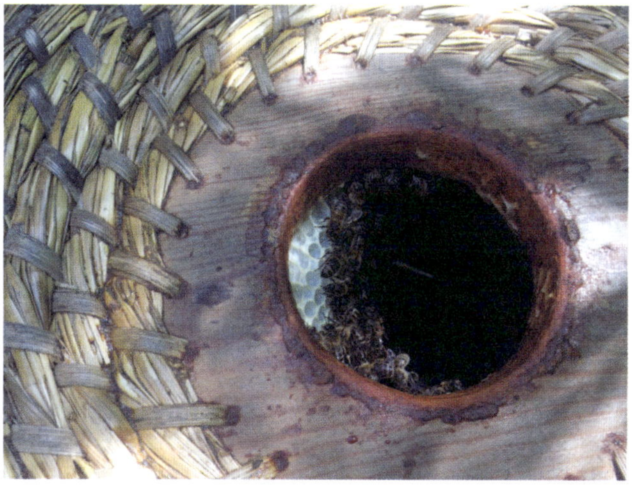

Top right: **Looking down into an occupied brood-skep made with a wooden ring.**

Above right: **A wooden lid fitted to the same brood-skep.**

Right: **A cap from a double-hiving situation, after the bees have been driven out (except for three scouts!)**

Chapter 8

Processing

Honeycomb

Once you have your honeycomb safely home in the kitchen, or the bee-shed, it is time to process it. From an apitherapeutic point of few, honey that's kept in its natural state in the wax combs is prized most highly. Often known as 'cut comb', this ought to be capped with wax on all sides. If you are wishing to sell or present it nicely, it ought to be of a suitable thickness to fit into the tubs or containers one may source from a beekeeping supplier.

Left **Pouring the honey harvest into jars.**

Below: **Cutting comb over a strainer.**

Below right: **Straining through a double-sieve.**

The second most highly prized honey is that which is drained or pressed. I go about this in a variety of ways, often cutting up the combs in the top of a straining tank, where the honey will drain through a sieve and settle in the bottom. The sooner you do this the better, when it is still warm. If you leave it too long, it may start to thicken and not run very fast. Once it has been strained, I will bottle it straight from the gate, into jars if it's for home use, and via an extra strain through a stainless-steel double-sieve if it's for market.

I have never tried to uncap wild comb and spin it, as one would spin a wired frame of honey from a super. I imagine it would be a messy job to hold and uncap wild comb, but I have been assured it can be done.

Collecting dry combs to render into wax.

Wax

If you have made a honey harvest, and perhaps some mead, the leftover wax can be rendered down and cleaned. More beeswax is harvested from a skep or fixed-comb situation than a conventional moveable frame hive.

It's a good idea to dedicate a large pan to your wax-rendering. Most kit you use will become waxy, as will the place you decide to do it and the clothes you'll wear, so protection is advised. Be careful of fire and naked flames; gas cookers are not advised as wax is very flammable.

Ideally you should clean the honey off your wax by washing with water. Many beekeepers like to use soft water, usually filtered rainwater or distilled water, because tap water can be alkaline and give the wax a dull appearance, which can be detrimental if one is making candles or wax products for show or sale.

Place the wax in your pan and add half a pint of water, then heat it up, stirring from time to time with a metal implement, trying not to boil it (**A**). The molten wax will float on top of the water, and impurities can be filtered out using a sieve. All of the hot water and molten wax will pour through the sieve, and it will catch what is often termed 'slumgum' (**B**). It is a mixture of old papery pupa casings and other debris, and once cooled it is very useful for fire lighting, makes a good nutrient compost, and can lure a swarm to a skep.

A

B

The cleaned-up remains can then be poured through the sieve into a suitable container and left to cool and set (**C**). The wax will set on top of the water, and in between the two there will be a fine layer of slumgum. Remove the loaf of wax from the container (**D**), discard the water and clean off the slumgum with a knife or a cheese grater (**E**).

Repeat the process with clean, soft water, heating it all up gently in the pan (**F**), then sieve again. For best results you can use a finer sieve, nylon, or a cloth filter. The more you repeat this process the cleaner your wax will become.

Once your wax is clean of debris (**G**), it's ready to heat up in a double-boiler, bain-marie, or electric wax-melter (**H**). Set up your preferred candle-moulds and pour in the wax when melted. Be careful: molten wax can splash and spill and get everywhere. Protect yourself and any clothes and suchlike that you wish to keep clean.

The wax has been poured into the table-candle moulds.

Opening the mould.

We burn most of our candles at home, and give the others as presents. After experimenting with candle dipping and many styles of mould, I've found the most luminous, long lasting and efficient are those of a simple table-candle design, about 240 mm (9.5 in) long with a wick of around 3 mm (0.12 in). Beeswax candles burn at a higher melting point than paraffin-wax, and therefore burn slower. This higher temperature also means they emit a brighter flame of a similar frequency to sunlight. They are non-toxic, smoke less, and emit that warm golden beeswax odour with esters and trace elements of propolis.

I heard a story that, years ago, wise old women buying candles at market would hunt for the darkest, dullest looking ones, knowing that they would last the longest.

A steady burn.

A skep illuminated by beeswax candles.

Chapter 9

Mead and Metheglin

Three cups a prudent person may take.
The first for constitutions sake.
The second for the sweetheart they love the best.
And the third to lull them to rest.

TRAD. ANON.

After extracting the honey, one may have a considerable amount of sticky wax leftover. This may be the beginning of a lifelong journey into the ancient craft of making mead.

Mead is often cited as the first alcoholic beverage, and over time many traditional varieties have evolved. Basic mead is simply made of honey and water. If one adds herbs and spices it is commonly known as a 'metheglin'. I love to make both, and craft many small, inconsistent batches that differ according to the season, the characteristics of the honey and botanicals that I might add.

One of the delightful discoveries I have made through skep beekeeping is the excellent quality of mead made using fixed-comb honeycombs. There is much more beebread, as well as other medicines of the hive, in the mix. These meads have a broader flavour, a more nutritious feel and many who have tasted them have remarked upon their fine character, expressing feelings of euphoria and satisfaction.

A good mead benefits from good honey. The best honey is of course that from one's own apiary. If you are a beekeeper, you may have some wax-cappings set aside after a honey harvest. If not, find a local beekeeper whose practices you favour and purchase three pounds to make a dry mead, or four to six pounds to make a sweet

A selection of meads. From left to right: sloe mead, self-heal metheglin, elderflower metheglin and cyser.

mead, depending on how sweet or strong you like it. You will have better results from honey that has not been heat treated, and many beekeepers do treat their honey in this way before bottling, to make it runny. This changes its characteristics and taste. My advice is to go for 'soft-set' honey, or honey that is labelled as 'raw' or 'unheated', which essentially means that it is unlikely to have been heat treated or homogenised after harvesting. Be aware that some of the honeys on the market are adulterated in one way or another. The type of honey you use will of course flavour your mead, giving you different tastes and textures, colours and characteristics. A mead made from a wildflower honey will be very different from a mead made from forest honey. A lime-flower honey mead will be much lighter than a chestnut or a heather-honey mead.

Honey bee collecting pollen from an elder tree (*Sambucus nigra*).

You may like to add a few herbs, the current trendy word for these being 'botanicals'. There are many to choose from. When brewing a mead, I generally head out to the gardens and hedgerows to see what is most vibrant. This may be ground ivy in April, elderflowers in June, or sloes in October. A great combination is elderflower, lemon, ginger, and cinnamon. Think herbal tea. Some people like to make a basic mead and then put a herbal tea bag in whilst ageing it. I find fresh herbs are more potent. Remember the adage, 'one man's bread is another man's poison'. Something like seaweed, which can be high in iodine, will be of benefit to many and harmful to some. If this is going to be your first attempt at mead making, set a steady course for a basic mead, with an innocuous handful of tried and tested culinary garden herbs and kitchen spices. If you are new to herbs, for a springtime batch I'd recommend a little fresh thyme, a handful of lemon balm and couple of fronds of ground ivy (also known as 'alehoof' or 'Jill-under-the-hedge'). Be cautious with strong herbs and overpowering spices, like rosemary and cloves. You may go along the route of adding herbs for their medicinal properties, or you may go via the route of creating sensory attributes and flavour tones. As a rule, tannins for flavour profile, entheogens for omniscient poesy and prophesy. Some people like to put black tea into the mix for tannins; I've never done so. Most traditional hops add a bitterness and an anti-microbial action. There were many that were commonly used before a particularly puritanical era when the Church banned the more stimulating varieties. After much mooting and deliberation, the religious stamp of approval was only given to the soporific variety.

Making mead

To make one gallon of mead or metheglin you'll need almost that amount of spring water, three to six pounds of honey, a handful of herbs, some beebread or bee-pollen, a lemon, a large pan (vat, cauldron, saucepan, or cooking pot), a five-litre demijohn or two, air-locks and bungs, a hydrometer, a hose or syphon, a straining bag or filter, a funnel, a long-handled wooden spoon, half-a-dozen clear wine bottles, corks, and a corking tool.

The following method is a for a wild-fermented mead, and the best results will come from using an authentic and raw honey.

1. Sterilise your equipment and rinse well.
2. Simmer herbs/botanicals/fruit for 5 minutes.
3. Turn off heat, stew for 20 minutes and strain once cooled.
4. Put the spring water, honey, tea, lemon and nutrients (ideally beebread) into the vat and mix.
5. Keep mixing until the honey dissolves, warming slightly, no higher than 30 degrees Celsius.
6. Check solution with a hydrometer. Adjust if needed.
7. Cover with a cloth and leave to stand in a warm or sunny place for a couple of days, stirring occasionally.
8. Filter, pour into a demijohn, top up to one 25 mm (1 in) below neck, and add airlock.
9. Set in full sun in the daytime, or a warm place.
10. Top up levels fully once the ferment slows a bit.
11. Rack off and filter after a couple of months.
12. When the water in the airlock becomes level on both sides, remove and put in a solid bung, racking off again if needed.
13. Age, rack off/filter. Bottle.

Bottled metheglin standing in front of different batches that are still fermenting in 1-gallon demijohns.

Keep it clean

Sterilise your equipment and rinse well. You may like to use a sterilising powder or Campden tablets for this, and then rinse thoroughly. If you are wanting to add fresh herbs to your batch, simmer them in a saucepan for five minutes, let them brew, allow to cool and then strain off the water into your mead-making pan, if it's not brewing in there already.

Once you've got your botanically brewed herbal tea solution, you can add the rest of the water and your desired amount of the raw honey. Start mixing and stirring with your long-handled wooden spoon. We call ours a magic wand. Over years of using the same spoon or spurtle for stirring the mead vat, it will accrue a certain amount of magic power, namely yeasts. It will only retain that power if it isn't put through a high-temperature wash or dishwasher. There's a famous story of a cheese-making company that upped their game and moved shop to a grand warehouse but couldn't recreate that distinctive tang their homemade cheese was renowned for. Someone suggested it was the new environment that was at fault, so they took down the farmhouse kitchen tiles and moved them to the new location. Lo and behold, that magic tang came back. It had been created and cultivated by the wild cultures in the home kitchen. The same can be said for your wild-magic mead-spoon, that will aid your wild-fermented mead, and become stronger over the years. Keep on stirring and agitating until most of the honey is dissolved. Honey is hygroscopic: it will continue to draw water into itself.

Citric acid

Citric acid is another magic ingredient. Add up to one quarter of a lemon or a similar quantity of another citrus fruit – lime or orange will suffice – avoiding the bitter pith. One might also use pineapple juice, although I never have. Any of these will help to regulate the acidity of your brew and be beneficial to the yeasts, aiding the vigour of the fermentation.

The recommended acidity for a mead is between 3pH and 4pH. There are many pH-testers on the market, from paper strips to digital meters. Very basically, if you wish to raise the pH, add water; and if you wish to lower it add fruit. A lot of the powdered citric acid on the market is derived from a genetically modified version of the fungus *Aspergillus niger*, but you wouldn't know it. Citric acid derived from this fungus and grown on a substrate of GMO corn is still considered a non-GMO. The choice is yours. Complaints of abdominal pain, cramps, bloating, diarrhoea, and nausea have sometimes been made due to overexposure to processed foods containing GMO-derived citric acid. A few citric acids are apparently still produced the old-fashioned way if you wish to track them down. If you don't have any citrus fruit to hand, you can add some apple for malic acid, which will serve the ferment in a similar way, yet with a different flavour.

Citric acid from lemons or other fruit helps with fermentation.

To boil or not to boil?

Many think it unnecessary to boil in this day and age, especially when wanting to retain the rich aromas of the honey. Even so, most mead making books will encourage you to do so, to create a biological blank canvas into which you add your chosen yeast. I certainly don't like to boil honey. The process changes its characteristics, dulls the flavour, diminishes its health benefits, and will kill off the wild yeasts that I want to keep for the wild ferment.

I think the practice of boiling honey in water might also be a vestige of times of less salubrious water sources than today. Good honey is acidic and anti-microbial enough to not need sanitising in this way. Any water should be simmered beforehand to be on the safe side, along with the botanicals of course to extract their properties before mixing with honey. After cooling this solution, I would mix this with the washings from the wax cappings left over from a honey harvest, often filtering it at the same time. Some herbs, though, may need to stay in the vessel during fermentation to make sure their desired oils are extracted. Examples here are sage, wormwood, and mugwort, if one desires a little bit of the chemical

Honeybee on early willow catkins with full pollen baskets (corbiculae).

thujone, the stress here being on the 'little bit'. Making metheglins can verge on herbal medicine, so please do your homework on safe and appropriate dosage and effects.

Beebread

For all basic meads, beebread is the most suitable nutrient. Often known by its Russian name *Perga*, and to Pliny the Elder as *Erithrace* and *Sandaraca*, it is pollen that has been stored in the hive. Some say it is the mythical ambrosia of ancient Greek mythology that is fed to the gods to keep them immortal. In many old bee-books, the fresh pollen is called *farina*, as it's like a powdery flour of tiny grains. Each individual grain has a hard outer wall like a nutcase called the endospore, which is made of sporopollenin. This hard polymer protects the 'seminal' power within, which can pass on the male genomic information, inspiring germination. These reproductive cells are known as gametes, and if you know any Greek, this 'g' word might conjure a few swear words and associated idioms. It is indeed fecund stuff. When worker bees collect pollen from flowers and catkins, they mix it with a little nectar to help shape it into a tiny loaf and carry it home in their pollen baskets, or corbiculae. It is then mixed with honey-bee enzymes and stored in waxen hexagonal cells, packed in layers, adjacent to or next to the bees' brood.

This becomes their main source of protein. Remember the ancient wasp ancestors of bees, who discovered this. Within the cells the pollen undergoes a mild lactic fermentation for around a week, which softens the outer endospore, making its inner nutrition more easily absorbed by the honey-bee larvae and the queen. If you aren't a beekeeper, it is worth asking a nearby one for a handful of beebread. Or you can find some online.

Stir in a tablespoon of beebread or bee-pollen; one loving spoonful will be ample to add to the warmed cauldron of mead must. Set it in a warm or sunny place and cover over with a cloth, tea-towel, or a piece of muslin. Then leave it to breathe for a couple of days, stirring from time to time. If the first rule of alchemy wasn't to never proclaim that one is an alchemist, then you might certainly be on your way to becoming one.

Hydrometer

A hydrometer is a very useful bit of kit to have. It's a weighted glass tube and will tell you your ratio of water to honey. In pure water it will sink deep, floating with the surface of the water level with the zero mark at the top of the tube (**A**). The more honey is added, the higher it will float. Keep putting honey into the mead must, making sure it is dissolved and well stirred, until the hydrometer floats up to the desired mark. There is a large, coloured section in the middle of the tube. This is blue in the pictures here; on others it may be a different colour that will have 'start wine' printed on it. When the hydrometer floats with the must level with the top of that mark, at 120 grams per litre, a dry mead is made. When it floats all the way up to the bottom of the mark, 240 grams per litre, a sweet mead is made (**B**).

The old-fashioned way of measuring the amount of honey in your mead was a fresh raw hen's egg. You would keep adding honey to the must until the egg floats to the surface with a part of it above the water about the size of one groat, which is roughly 25 mm (1 in) (C).

Straining and starting

Any wax will float to the top. This and any bits of fruit or herbs floating around can be happily left in the mead must and will be easily removed by filtering through your straining bag when it's ready to put into your demijohn. Save the precious beeswax to wash and add to your collection of beeswax to process later to clean and render for candles, creams, balms, and more. When straining, avoid metal sieves and metal spoons, as they can taint the mead.

On day two, when you stir your mead, you may see some froth on the surface and hear a slight fizz; this is a great sign. If you have another suitable vessel, you can also aerate and invigorate the must by decanting it from one to the other a few times. If you aren't getting that fizz, set it in a sunny place, or warm it slightly to around 30 degrees Celsius, stirring occasionally, and wait another day before putting it into your demijohn.

I use a fine-meshed nylon bag inside a funnel to filter out any last bits as I pour the liquid into the demijohn. It's worth leaving an air

Must and yeast

The 'must' is a term for the unfermented solution: the mixture of ingredients to which you might add a yeast if you wish to. For many years now, I have been successfully crafting wild fermented meads and metheglins without adding a yeast. This can be achieved simply by allowing the musts to breathe a while, accruing more yeasts from the atmosphere, and the oxygen they need to launch them into action. To make a more conventional mead, one might quickly boil or pasteurise the must to remove any scum and inhibit any existing yeasts, and then add some trusty, standardised white wine or sparkling wine yeast from a packet.

gap of about 25 mm (1 in) in the top, both to aid the fermentation and to avoid the solution frothing its way up through the airlock during the initial aerobic fermentation, which can be quite vigorous once it gets going. I use the standard water-bubbler airlocks; there are various kinds to choose from.

Demijohn and airlock

Once you've got your mead must into your demijohn (often called a carboy in the US), put on the airlock, and set it in full sun during the day if possible. This can be on a windowsill or outside, but if it's outside, make sure to bring it in to keep it warm at night. To wake up the yeasts, I will often give the whole thing a shake from time to time if it is sluggish. Soon enough your airlock will start bubbling quickly, between a few days and a week later. Be patient and keep it warm. Sometimes it can get going at around 80 beats per minute at the beginning, and then gradually slow down after few weeks. Once it has slowed down, and there is no risk of the solution bubbling up through

Above: **Demijohn with an airlock.**

Right: **The lees.**

the air lock, you can fill it up to the top by adding some water, or a mix of water and honey. Then, set it somewhere slightly cooler and let it bubble slowly for a few months, during which time it will start to shed a sediment at the bottom of the vessel. This is called the 'lees' and is mostly dead yeast, and often pollen in my case if I'm making a special pollen-rich batch.

Racking off

Once your mead has shed its first lees, you can 'rack it off'. This is a wine-making term for siphoning it into a clean demijohn, leaving the lees behind. When you put your syphon tube into the mead, try not to disturb the sediment at the bottom, and begin siphoning. Racking off can help clarify your mead. You'll get to taste a bit of your batch if you are using a simple siphoning tube. Make sure you top up the liquid levels in the new demijohn, to minimise any oxidation, which can create a bad taste. You can do this with spring water, or with a mix of water and honey as before if you wish to revitalise the fermentation, sweeten the mead, or make it stronger. Remember you can add herbs at any time. Sometimes I will gather meadowsweet (*Filipendula ulmaria*) whilst fetching water from the springs here at Westmill Farm, and I'll happily add that straight into the demijohn of a newly racked mead. I'll do the same thing with borage (*Borago officinalis*).

When your mead is racked, clean and replace the airlock as soon as possible. It's a good idea to bung the demijohn whilst cleaning and refreshing an airlock,

Racking off.

Honeybee and ground ivy (*Glechoma hederacea*), also known as 'alehoof' or 'Jill-under-the-hedge'.

to stop those little *Drosophilia* fruit flies getting in. My grandfather called them the 'vinegar fly' because if they do get at your mead or wine, they will turn it sour, and it'll be on its way to becoming an artisan vinegar. Which is all well and good, and there is evidence from Pliny that honey vinegars were prized above wine vinegars, but this is not our desired intention. These flies often manage to make their way into the water well of the bubbling air locks too, which may well need topping up at some point due to evaporation. Any equipment that they manage to get into or onto will need a thorough clean.

There will come a day when you check your mead, and it will bubble once every five minutes, and then eventually once a day; perhaps in a cooler season, you'll notice the airlock level out. Be aware, though, that it can level out in the cool of the evening and start bubbling again on a hot day. If it stays level, then it's time to rack off any sediment again, top up your liquid level, and you can then plug it with a solid bung and leave it to age for a few months before bottling. Alternatively, you may bottle it straight away if you want to, if you think it's not going to shed another lees or start fermenting again. Some brewers like to be sure to stop the ferment and reduce the risk of oxidation here by adding sulphites or half a crushed Campden tablet to the gallon. If this is unappealing and your airlock stays level even on a warm day, I would bottle it. There is a slight risk that it can start to ferment again, especially if it warms up during a really hot summer. There may be a slow anaerobic fermentation occurring after the initial rapid aerobic fermentation. Putting the demijohn outside for a few cold February evenings will completely stop the fermentation, and can aid clarification too.

Filtration

I inherited a few filters and filter powders from my grandfather, and there are many gadgets on the market to help with filtration and clarification if your desired outcome is a crystal-clear bottle of mead. I'm happy either way; it's always nice to open a clear mead that effervesces slightly when you pour it into a glass, and it's also great to sup that slightly cloudy, pollen-rich batch. The only filter I tend to use today is a fine nylon mesh sieve shown in the racking off photos. Some traditional hops are good for clarifying a brew, like ground ivy. Be patient, and let the lees settle. Some batches you may rack off three to four times, and as if overnight, one day in the middle of winter a demijohn full of mead will suddenly clear.

Bottling

When bottling, use corks and not screw-tops or metal caps. These can be normal wine corks if you're intending to leave it in the bottle for a while, and it really does get better the longer you leave it. Or you can use the plastic-topped corks that can be removed by hand if you think it'll be drunk sooner rather than later, or if you are wanting to enter your best mead into a honey show. There are various corking machines on the market, and various bottles. I recommend getting your corks and bottles from the same supplier, as sometimes they can be slightly mismatched, the bottles being a bit too tight or the corks too big. Once bottled, store it in a cool dark place. If you're not too fussy about filtration, like me, then you may find a tiny bit of sediment in the bottom of a bottle after a while, similar to what you see with some ciders. If you are wanting to produce a gin-clear mead, then invest in a good filtration system and apply due patience.

Once you get started the possibilities are almost endless, and the longer you can leave it corked in the bottle, the smoother it gets. Think of the archetypal one-year-and-one-day, as a good rule for ageing. I remember making a batch in a five-litre container many years ago and had written it off because the airlock got knocked off, some of the contents had spilled, and I had thought it spoiled. I sighed, bunged it, and put it at the back of a shed. There it stayed for several years until I needed the container. It was a large plastic one with a tap fitted. One day I unearthed it, and out of curiosity turned the tap. To my surprise it was a clear and dark mead with treacle tones and fine tannins, as if oak-aged. It had a medium-sweet and nourishing taste, full of clarity and cognisance, a touch of Hungarian Tokaj, with woody and pleasant undertones of propolis, cherry, and toffee. I quickly bottled it all, and even though I tried my hardest to put a bottle aside for the future, it was just the best treat to splash into the glass of any visitor, give away as birthday gifts, and make the odd libation with here and there. Within a couple of months, it was all gone.

Metheglins

As you have learnt, a mead made with herbs or spices is a called a metheglin. For an early summer batch, you could try a generous number of elderflowers, picked on a sunny day and removed from their bitter stalks. Perhaps add a scattering of dogrose petals and a frond or two of St John's Wort. Then put in a teaspoon of cinnamon and ginger. For late summer, both borage and meadowsweet are excellent. A little later you might like to make a more astringent batch with some fresh hops.

Bee-line for the borage (*Borago officinalis*).

Borage and St John's Wort bubbling in the demijohn.

If I'm mead making in autumn, I'll certainly put a few sloes, some slices of apple and autumn raspberries into the must. Sloes are the small dark fruit of the blackthorn tree, thought to be the wild ancestor of European and western Asian plums. They contain citric and malic acid, as well as tannins. Gather a basket or a bag full of them from September onwards, ideally from a hedgerow on an organic farm, far from a roadside. Be careful of the thorns, which can cause inflammation and sometimes infection. Bring the fruit home, cleaning off any leaves and twigs. Just like gathering them for sloe gin, if there hasn't yet been a good frost to split or soften them, you can prick them with a fork or put them in the freezer for an hour. This will enable their flavour to flood out. Alternatively, one can simmer them in some water, mashing them from time to time until the liquid is infused with their richness. This will kill off some of the beneficial yeasts, so why not throw a few fresh ones in after it has cooled a bit. Filter out the stones and then it's ready to add to your mead vat.

Ingredients can be added or mixed in to your mead at any stage of the process, but not easily removed. Balancing the botanicals and the honey is half the craft. Some of the most notable meads and country wines I've made were largely flavoured or aided by sloes. Some might call a mead that has fruit in a 'melomel'. Strictly speaking though, a true melomel is made with no water, only fruit juice and honey. If made with apple juice and honey, it's known as a 'cyser'. If I put sloes into a metheglin, you would correctly call that a fruit metheglin. Indeed, a basic mead with some plums or sloes added is a 'fruit mead' and not a melomel.

Seaweed metheglin

Detail of sunlit bladderwrack.

A few years ago, I was given a couple of bags full of seaweed from west Wales. It was fresh and nutritious bladderwrack destined for the garden as a rich mulch. There was a good weight of the plant, high in vitamins and minerals. I opened the bags and they released that rich, salty aroma of the seashore. Fetching a spade, I emptied some seaweed onto the ground and chopped it up into a mulch to feed a Sutherland Kale patch.

Soon after, my mind travelled to the Scottish coastal traditions of making seaweed beds for planting grain, which in turn imbued their malts and beers with unique phenotypes full of the rich terroir of the land, sea, and sky there.

An intuitive impulse took over as I saved one bag of the wrack for experimenting with the mead vat. Waiting indoors here, in the Vale of the White Horse, far from the sea, were all the ingredients for a few batches of mead. I had some fresh wax-cappings to hand from a recent honey harvest and some demijohns full of good spring water.

I knew that the seaweed Irish Moss, amongst others that contain carrageen, had long been used as a fining for clarifying brews. I learnt that the original kombuchas may have been fermented from a kelp tea, and that a once-popular German beer called Gose draws its salty character from seawater. A good mead really benefits from a nutrient that not only broadens the taste, palate, and feeling signatures, but encourages the yeasts to work more vigorously.

I went into the kitchen and mixed the wax-cappings and the spring water together into a large fermenting bucket, stirring and singing shanties as the wax began to float to the top of the liquid. Delving down with my large wooden mead-making spoon, I squashed and dissolved any lumpy bits of honey there. Then I made a cuppa and scoured the bookshelves for more information on bladderwrack.

After a little more research about brewing with seaweed online, I washed it quite a few times in cold water to remove the over-saltiness and began to chop it up, to break it down, to add to the mead must.

The must was filtered into demijohns and set in the sun to warm. It soon shed a sediment and was racked off about a month later. I left the foggy brew to age for a few more months and then bottled it in due course. Opening the first bottle was an event! To our amazement, when the cork was popped it foamed exquisitely. Poetically, a wavelet of affirmation lapped at the shores of our anticipation. Sloshing into the mead horn it echoed the briny flood of Venus and contained all the character of the ocean. I wished I'd made a second batch.

Your field of experimentation as a mead-maker is almost endless. I make many small batches, each one different. If you decide to set up a meadery and commercially produce a reliably similar product each time, then standardisation of every detail of your operation is important.

Above: **Wassail bowl, shown in the *Illustrated London News*, 1842.**

Drinking vessels

Horns

Once aged, bottled, and finally opened, there are many significant vessels from which to sup and share mead. A large drinking horn is not only a status symbol, but also denotes generosity, hospitality and abundance. We often associate drinking horns with Vikings, Celts and Anglo-Saxons, but their use was widespread. They were commonplace in Ancient Greece, Thrace, and Rome. We see them in African rituals and Asian folk-customs today. This natural vessel has been the perfect cup since time immemorial. In the later classical eras it is echoed in the 'horn of plenty', or cornucopia, which has numerous mythical beginnings.

Below: **Carved mead horn.**

Communal drinking

Amongst other drinking vessels needing mention are the 'mazer bowl' and the 'wassail bowl'. Both are broad and shallow and are passed around to supply many. The deeper, four-sided, and usually four-handled 'mether cup' is also used for communal drinking, affirming bonds of friendship. Originally, these were made of wood, and later of ornate silverware. A famous silver mether cup is used as the All-Ireland Senior Hurling Championship trophy, the Liam MacCarthy Cup. You must drink from a corner of a mether, as supping from a flat side will get you rather wet and sticky.

Mead horns can be similarly generous, especially the larger ones. When you raise a large communal horn full of mead up to drink, you can't always see the elixir flowing towards you, and you still can't see it, and then

A selection of sacred vessels: multi-handled friendship cups, mazer bowls, mead horns, caudle cups, and more.

suddenly you're wearing it! My advice is to go slow and steady, and lower the horn down carefully after drinking before you pass it on, as it can splash back if you move it too quickly.

The adage that wood warms you twice, first when cutting and chopping it, and second when fuelling the fire springs to mind; mead can inspire you twice, first whilst making it, and again when drinking and sharing. The large horns, various mead bowls, and the two-, three-, and four-handled cups are designed to be passed around, affirming bonds and trust. We are all aware of the perils of overindulgence but they may well have served as a kind of inoculation within one's community, like an ancient version of the flu jab.

The End...

This book began with apian creation myths from the most ancient cultures, and to end with some kind of deluge or destruction myth might seem pessimistic. Einstein of course warned that we wouldn't have long left if bees disappeared, but I feel sure bees will still be here long after we've gone.

There is no single 'bee-saving' silver bullet, but considering their role as bio-indicators, it may be sensible to extend our view of bee-health to include the fields and hedgerows, gardens and forests, moors and riverbanks. Bees have been looking after each other for a vast amount of time, and what they need, just like most creatures, is a healthy environment and a good diet. If we enrich, protect, and preserve organic lands, gardens, waterways, and wildlife, they'll continue to thrive by themselves. To give them the opportunity to gather salubrious substances, benefiting from a biodiverse and organic environment as free from toxins as possible, is paramount.

Skep making and skep beekeeping have been rewarding research avenues into the old sustainable ways, and keeping these ancient technologies alive is not just a hobby, a heritage, and a hearty lifestyle; it may well be the future of beekeeping one day. So I would like to push three main points here:

1. You can sustainably ensure healthy bee colonies by keeping them in skeps, and make an ethical harvest without harming them.
2. Most human-managed honey bees in the world are kept in naturally made, locally sourced vessels, and live on fixed-comb.
3. Bees kept in such a way, on fixed-comb, that are freely allowed to swarm and never treated with chemicals, are the healthiest bees in the world.

I would add to this list the resilience and wellbeing of honey bees in this situation, the fine quality and characteristics of the honeycomb harvested from a skep, the fulfilling nature of the craft of skep beekeeping, and the perpetual prophylactic

Some midsummer botanicals amidst an array of skep products and tools.

Wassail! Slainte! Lachyd da!

privileges of beekeeping in general. Then add, again, the deepening connection one will develop to the seasons, trees, and plants, with an enhanced natural awareness of one's wider environment, garden, and the life-affirming rhythms of nature, within and without.

You may of course wish to keep a few colonies of bees in skeps and not harvest any honey or wax, and simply provide a habitat for wild bees, forsaking whatever they might leave behind for moths and mice, birds, and beetles. I meet more and more people who are simply providing places for honey bees to live unhindered in skeps and log-hives.

Whichever way you wish to proceed, if you do, you are the future of beekeeping, and I wish you and yours, and the bees you may meet along the way, all the very best. Whatever the future may hold, we know that skep making, skep beekeeping, honey harvesting and mead making are exceptionally rewarding, time-honoured and fortifying arts reaching far beyond ourselves, here and now.

ACKNOWLEDGEMENTS

I was lucky enough to meet the British beekeeping historian, book collector, and writer Karl Showler, who encouraged me along the route of the skeppist. Likewise Martin Buckle, Jim Ryan from Ireland, and David Chubb have all been inspiring. I also feel fortunate to have been befriended and guided by the late David Charles of Somerset, the late Bill Mundy of Dartford, and Bronwen and David Wright of Edinburgh. They are all beekeepers who have delved into skeps, helping to perpetuate their charms. Nicola Bradbear and Bees for Development have become good friends, expanding my nerdy beekeeping interests far beyond the shores of the UK and Ireland. Beekeeping historian Will Messenger has also been a good friend and fellow member of various beekeeping history groups, alongside Norman Carreck, the late Peter Tomkins, and Pam Hunter. Through these groups I have collected some fascinating stories over the years, including numerous anecdotes of fathers and grandfathers who settled bees into baskets and kept the dwindling craft going. Two friends needing mention are J.J. Middleway, who helped initiate my first steps into beekeeping as we adventured into star lore, and Fanny Soirat, whose insight and wisdom helped temper my wit and words. I'd also like to acknowledge Verity Sharp and Patrick Randall who invited me to co-host the Living Beeing podcast, where we interviewed a few of those mentioned above, and many more, including myself on an episode dedicated to mead making.

Karl Showler had worked for the International Bee Research Association (IBRA) which was founded by Eva Crane in 1949. Two of her books, *The Archaeology of Beekeeping* (1993) and *The World History of Beekeeping and Honey Hunting* (1999) are highly recommended and provide much insight into fixed-comb beekeeping.

Lipwork api-sculptural project in the author's workshop.

GLOSSARY

Apian: of or relating to bees.

Apitherapy: medicine that uses bee products to treat illnesses and symptoms.

Aril: scarlet flesh around a ripe yew berry.

Barm: yeast, taken from the foam or scum off the top of a fermenting liquid. Used to start a ferment in a liquor, bottle condition an ale, or leaven a bread.

Beebread: pollen that has been collected by bees, mildly fermented and packed into wax combs adjacent to their brood. It is a probiotic protein for the larvae.

Bennets: grasses, rushes, and suchlike, used as a lipwork material.

Cloom / cloome: a composite material for plastering skeps and alvearies. Often containing cow dung, sand, wood-ash, and chopped fibres like straw, flax, or horsehair.

Coppe: a small wooden disk with a hole in it, set into the ceiling of a skep.

Copse: an area of coppiced woodland.

Corbiculae: pollen-baskets, or panniers, on the hind-legs of worker bees.

Diploid: containing two sets of chromosomes.

Eke: a ring of lipwork basketry with no top or bottom, often made from old skeps, usually four rowls deep. They are placed underneath a skep when it is nearing fullness to give the colony more room, and are taken off at harvest time.

Fid: a small, pointed tool, like an awl, but with a channel running through it. Used for rope splicing, and to penetrate a material to create space to thread another through, be it lapping, rope, or willow etc.

Fixed-comb: wild comb attached to the ceiling and walls of a skep, or other vessel. As in 'fixed-comb beekeeping'.

Gart: a hoop to set onto a hackle, usually made of wood or metal.

Garth: a garden, from the Welsh gardd.

Hackle: conical straw cap for a skep, to protect it from the elements. Also known as a 'coppet'.

Haploid: containing a single set of chromosomes.

Hygroscopic: moisture-absorbing.

Lipwork: a style of straw craft, being bound with lapping.

Lissom: a twisted length of straw used for binding.

Melissean: of bees. From the Greek *melissa*, for bee; also giving the Latin for the European bee, *Apis mellifera*.

Melissomelos: 'The Bees Madrigal', written and composed by the Revd Charles Butler.

Metheglin: a mead made with herbs and / or spices. Used as a folk-medicine and a tonic.

Nasonov pheremone: communication scent produced by bees as an assembly message.

Neat's dung: cow dung.

Perga: another name for 'beebread'.

Phytological: botanical, plant-based.

Prophylactic: a medicine or course of action used to prevent disease.

Psychopomp: a deity or spirit who guides the soul to an afterlife.

Rowl: roll of straw; an average skep has about 16 rowls from top to bottom.

Ruskie: a Scottish term for a skep or basket.

Skep: a lipwork basket.

Skeppist: a skep beekeeper.

Spleet: a split piece of wood, sharpened and driven into a skep to help support the wax combs.

Spurtle: a Scottish wooden stirring utensil.

Strawne: made of straw.

Wassail!: Old English drinking pledge meaning 'be healthy', and replied to with 'Drinkhail!'

BIBLIOGRAPHY

Alston, F. *Skeps, Their History Making and Use*. 1987. Mytholmroyd. Northern Bee Books.

Alston, F. *Hives and Honeybees in Signs and Symbols*. 1998. Mytholmroyd. Northen Bee Books.

Beck, B.F. & Smedley, D. *Honey and Your Health*. 1971. New York. Bantam.

Beswick, F. *Traditional British Honey Drinks*. 2007. Herefordshire. BBNO.

Bradbear, N. *Plant Trees, Sow Seeds, Save The Bees*. 2021. London. Penguin.

Brecklemans, T. *Korfvlechten*. 1979. Amsterdam. Cantecleer.

Breverton, T. *The Physicians of Myddfai*. 1988. Carmarthen. Cambria.

Brown, Ron. *Look, Listen, Learn at the Hive-Side*. 1994. Torquay. Self-published, printed by A&D printers.

Brown, Royden. *How to Live to the Millenium – The Bee Pollen Bible*. 1989. Arizona. Hohm Press.

Brown, Royden. *The Beehive Product Bible*. 1993. New York. Avery Publishing Group.

Buchmann, S.L. & Repplier, B. *Letters from the Hive: An Intimate History of Bees, Honey, and Humankind*. 2005. New York. Bantam.

Buhner, S.H. *Sacred and Herbal Healing Beers*. 1998. Colorado. Brewers Publications.

Butler, C. *The Feminine Monarchie*. Second Edition. 1623. (facsimile 1985. Mytholmroyd. Northern Bee Books.)

Columella, L.J.M. *De Re Rustica, Book IX*. Trans. Boyd Ash, H. Ed. Moyer, R. Great Britain. Amazon.

Crane, E. *The Archaeology of Beekeeping*. 1983. London. IBRA.

Crane, E. *The World History of Beekeeping and Honey Hunting*. 1999. Oxford. Routledge.

Crane, E. & Graham, A.J. *Beehives of the Ancient World*. 1985. Cardiff. IBRA.

Digbie, K.K. *The Closet of Sir Kenelme Digbie Kt. Opened*. 1669. (facsimile 1967. Mallinkcrodt Chemical Works.)

Dixon, L. *Bees and Honey: Myth, Folklore and Traditions*. 2013. Mytholmroyd. Northern Bee Books.

Edwards, T. *The Lore of the Honey-bee*. 1912. London. Methuen.

Fearnley, J. *Bee Propolis, Natural Healing from the Hive*. 2001. London. Souvenir Press.

Ferry, D. *The Georgics of Virgil: A Translation*. 2005. New York. FSG Books.

Foster, A.M. *Bee Boles and Bee Houses*. 1988. Oxford. Shire Publications.

Fraser, M. *Beekeeping in Antiquity*. 1952. London. University of London Press.

Galton, D. *Survey of a Thousand Years of Beekeeping in Russia*. 1971. Gerrards Cross. BRA.

Galton, D. *The Bee-Hive: An Enquiry into its Origins and History*. 1982. Hebden Bridge. Northern Bee Books.

Gayre, G.R. *Wassail! In Mazers of Mead*. 1948. London. Phillimore & Co.

Gendrolis, A. *Propolis, Natural Medicine*. Trans. Žiukienė, I. 2017. Lithuania. JSC.

Hoffman, D. *Welsh Herbal Medicine*. 2007. Aberteifi. Abercastle Publications.

Horn, T. *Bees in America: How Honeybees Shaped a Nation*. 2005. Kentucky. University Press of Kentucky.

Ioyrish, N. *Bees and People*. 1974. Moscow. MIR.

Ioyrish, N. *Curative Properties of Honey & Bee Venom*. 1977. Hawaii. University Press of the Pacific.

Isaac, Rev. J. *The General Apiarian*. 1799. Exeter.

Kaal, J. *Natural Medicine from Honey Bees*. 2017. Mytholmroyd. Northern Bee Books.

Keys, J. *The Practical Bee-Master*. 1780. London.

Mollan, C. (Ed.). *Reprinted Papers on Bee Husbandry*. 1980. Tallaght. Royal Dublin Society.

Ogden, R.B. *In Pursuit of Liquid Gold*. 2001. Cornwall. BBNO.

Payne, J.H. *The Beekeeper's Guide*. 4th edn. 1842. London. Newby.

Petry, M. and Shavkam, A. *The Bear and the Bee: Traditional Medicinal Uses of Honey in Russia*. 2016. UK. Amazon.

Pettigrew, A. *The Handy Book of Bees*. 1870. (Reprinted 2009. Milton Keynes. Lightning Source.)

Ramírez, J.A. *The Beehive Metaphor*. 2000. London. Reaktion Books.

Ransome, H.M. *The Sacred Bee*. 1937. London. Butler and Tanner.

Seeley, T. *Honeybee Democracy*. 2010. Princeton University Press.

Showler, K. *Some Hints on the Art and Mystery of Skep Making*. 1990. Mytholmroyd. Northern Bee Books.

Snelgrove, L.E. *Roman Bee Lore*. 2023. Mytholmroyd. Northern Bee Books.

Southerne, E.A. *Treatise Concerning the Right Use and Ordering of Bees*. 1593. London. Thomas Woodcocke.

Spiller, J. *The Advantages of the House Apiary*. 2013. Mytholmroyd. Northern Bee Books.

Storch, H. *At the Hive Entrance*. Trans. Celis, F. 2010. Brussels. European Apicultural Editions.

Taylor, H. *The Bee-keepers Manual*. 6th edn. 1860. London. Groombridge & Sons.

Thorley, J. *Melisselogia. Or, The Female Monarchy*. 1744. London.

Tusser, T. *Five Hundred Pointes of Good Husbandrie*. 1573 & 1577. London.

Vernon, F. *Hogs at The Honeypot*. 1981. Burrowbridge. BBNO.

Walker, P. *Hive Enclosures for Bee-skeps in England*. 2011. British Association for Local History.

Wellcome, H. *Ancient Cymric Medicine*. 1988. Edmonds, WA. Sure Fire Press.

Wildman, T. *Treatise on the Management of Bees*. 1768.

Useful links

www.apitherapy.com – leading apitherapy hub, courses, workshops, resources, and products.

www.bbka.org.uk/Pages/Category/about-spring-convention – start of season trade-show with classes and lectures.

www.beesfordevelopment.org – beekeeping charity encouraging nature-based beekeeping worldwide.

www.honeyap.org – the Honey Authenticity Network

www.honeyshow.co.uk – end of season beekeeping event with show, classes and lectures.

www.livingbeeing.com – podcast interviewing people who are passionate about bees, from entomologists to botanists, scientists, beekeepers, honey experts, historians, artists and api-therapists.

INDEX